Nutrition of the Angiosperm Embryo

RESEARCH STUDIES IN BOTANY AND RELATED APPLIED FIELDS

Series Editor: **Dr. P. S. Nutman, F.R.S.**

1. Root Nodules of Legumes: Structure and Functions
   F. J. Bergersen

2. Acid Soil and Acid Rain
   The Impact on the Environment of Nitrogen and Sulphur Cycling
   I. R. Kennedy

3. Biochemistry of Virus-Infected Plants
   R. S. S. Fraser

4. Nutrition of the Angiosperm Embryo
   David R. Murray

Nutrition of the Angiosperm Embryo

David R. Murray

RESEARCH STUDIES PRESS LTD.
Taunton, Somerset, England

JOHN WILEY & SONS INC.
New York · Chichester · Toronto · Brisbane · Singapore

RESEARCH STUDIES PRESS LTD.
24 Belvedere Road, Taunton, Somerset, England TA1 1HD

Marketing and Distribution:

Australia, New Zealand, South-east Asia:
Jacaranda-Wiley Ltd., Jacaranda Press
JOHN WILEY & SONS INC.
GPO Box 859, Brisbane, Queensland 4001, Australia

Canada:
JOHN WILEY & SONS CANADA LIMITED
22 Worcester Road, Rexdale, Ontario, Canada

Europe, Africa:
JOHN WILEY & SONS LIMITED
Baffins Lane, Chichester, West Sussex, England

North and South America and the rest of the world:
JOHN WILEY & SONS INC.
605 Third Avenue, New York, NY 10158, USA

Library of Congress Cataloging in Publication Data

Murray, David R. (David Ronald)
Nutrition of the angiosperm embryo / David R. Murray.
p. cm. — (Research studies in botany and related applied fields ; 4)
Bibliography: p.
Includes index.
ISBN 0 471 92153 X (Wiley)
1. Angiosperms—Embryos—Nutrition. I. Title. II. Series.
QK867.M95 1988
582'.130433—dc19

British Library Cataloguing in Publication Data

Murray, David R. *1943–*
Nutrition of the angiosperm embryo.
1. Flowering plants. Embryos
I. Title II. Series
582.13'0433

ISBN 0 86380 077 7
ISBN 0 471 92153 X (John Wiley & Sons Inc)

ISBN 0 86380 077 7 (Research Studies Press Ltd.)
ISBN 0 471 92153 X (John Wiley & Sons Inc.)

Printed in Great Britain by SRP Ltd., Exeter

Contents

LIST OF TABLES:

Preface

We depend upon plants for own own survival. In a world where plant species are increasingly threatened with extinction, and where sound agricultural and conservation practices are often disrupted by thoughtless exploitation, a deeper understanding of the requirements of plants is clearly needed.

Multicellular organisms such as higher plants face problems of nutrient uptake, transport, assimilation and allocation that are not shared by single-celled organisms. Because of their inherent complexity, higher plants are more challenging and more rewarding as research subjects than the latter. However, since many tertiary institutions in Western countries provide overly narrow specialization, far too many plant scientists do not appreciate this complexity. This often leads to unnecessary and sometimes serious errors of interpretation.

Let me recall statements on this problem from two emin ent 'holistic' plant scientists. First, a warning from C. W. Wardlaw (1952): " . . . the overemphasis of any single aspect, while the whole is not kept in proper perspective, will almost certainly lead to the fabrication of unstable theoretical superstructures, destined to crumble . . ." Secondly, a statement from M. Evanari (1985): "Since the whole (plant) is more than the sum of its parts, it can be comprehended only by considering all the structures and functions in their totality and their mutual interplay".

In this book I have endeavoured to comprehend the whole plant, in the way recommended by Evanari. This is not to say that my interpretations are free from error, and I will welcome correspondence on any topic. I would like to express my appreciation to everyone who supplied information and material, especially Mark Peoples, and to holders of copyright, who gave permission for the reproduction of items already published.

It is a pleasure also to thank the Series Editor, Phillip Nutman, for helpful comments on drafts of the text, and Ivan Kennedy, for specific comments on Chapter 2. I am grateful to the staff of the Centre for Teaching Development, The University of Wollongong, for help with the preparation of Figures.

David R. Murray

CHAPTER 1
Introduction

An appreciation of the way an embryo sporophyte is nourished inside a developing seed begins with an attempt to understand the ways in which the maternal plant gathers, transforms, transports and utilizes its essential nutrients. Many questions may be posed about the nutrition of any particular species of plant. Which of the compounds or elemental ions available in the aerial and edaphic environments are essential or useful nutrients, which are of no consequence, and which are potential toxins? At what concentrations might essential substances become detrimental or even lethal? How does a plant cope with an imbalance of either anions or cations from the soil solution? How are nutrients or their metabolic derivatives transferred from one part of the plant to another? How might we best study the way priorities are developed by individual plant organs in acquiring nutrients, and in particular, how might we use this knowledge to improve the productivity of the relatively few species that we depend upon for food?

In the context of plant-soil relationships, some of these questions have recently been addressed in the book in this series entitled "Acid Soil and Acid Rain - The Impact on the Environment of Nitrogen and Sulphur Cycling" by Ivan R. Kennedy (Research Studies Press, 1986). Others will be considered in relation to the main theme of this book. A brief formal account of plant nutrition (Chapter

2) is followed by an outline of the salient features of carbon and nitrogen assimilation (Chapter 3). The functioning of vascular tissues in the translocation of nutrients is also of fundamental importance. A number of the classic studies on phloem function have involved studies of fruit development. We now have abundant evidence that the phloem provides the principal route for nutrient transfer from supply organs to developing fruits and seeds (Chapters 4 and 5).

The pathways that individual nutrients follow from phloem sieve elements to the tissues of the fruit or to seedcoat secretory cells are complex, and at present controversial. A substantial portion of this book is devoted to considering the mechanisms of nutrient transmission to the embryo sac, and the integration of fruit and seed development (Chapters 6 and 7). The recently devised technique of removing the embryos from seeds still attached within developing fruits permits the analysis of nutrients and growth substances actually secreted by the seedcoats - and provides a useful way of assessing how the nutrient supplies from the maternal plant change during development (Chapters 6 and 7).

Modification and storage of nutrients within the embryo are topics dealt with only briefly, as they have been thoroughly reviewed elsewhere. Strictly, neither embryogenesis by somatic means, nor *in vitro* embryo culture, are in themselves seed development. Yet these topics concern the propagation of plants under conditions that generate new genotypes by somaclonal variation, which is far more effective than conventional breeding techniques depending only on a sexual cycle. The development of media suitable for culturing embryos "*ex utero*" is described in detail (Chapter 8). Very few attempts have been made so far to relate the composition of such media to the complex variations in nutrient supply that are inherent in the development of the embryo of any given species. The underlying causes of somaclonal variation (Chapter 9) probably relate to some kind of nutrient stress - but no systematic study of the nutrition of somatic embryos has yet been carried out. Despite (or because of?) the arbitrary nature of media recipes, somatic embryogenesis has now been successful with enough species to indicate its general potential as a means of improving crop species.

Priority areas for research have been indicated throughout this book, which hopefully will kindle the interest and enterprise of young researchers. There is enormous scope for further investigation - and little time left for wasted effort or misguided allocation of research funds. The predominance of higher plants as human food sources renders urgent the task of elucidating the processes whereby nutrients are distributed to those parts of the plant that we can safely harvest and eat, and protecting these plants from pests and diseases with due regard for the environment. Plants with genetically endowed resistance to the toxins of pathogens should need fewer pesticide applications, and eventually, more specific ones.

The areas of arable land and fertile soil available to us are dwindling daily at an alarming rate. If the plants we grow in the areas we have are able to perform better, or we lose fewer of them prior to harvest, then yields can be sustained while educated measures to stem population growth are given time to take effect.

Finally, some comments are necessary on the conventions adopted to streamline communication. S. I. units have been used, except when these seem inappropriate. Concentrations are usually given as molarity (M, mM or μM), or sometimes amounts are listed in a convenient form for preparing solutions, as in the original publications. Enzyme activities are usually expressed as units, where the unit is 1 μmol of substrate converted to a product or products per minute under the conditions of assay.

For common metabolites, it is understood that amino acids and amides are in the L- form, and sugars, organic acids and their phosphate esters are in the D- form. Abbreviations are standard, and are as listed in the journal 'Plant Physiology', e.g. MW for molecular weight, and Tris for tris(hydroxymethyl)aminomethane.

Common names for plants are frequently used, but in the index, these will lead back to botanical names, to avoid ambiguity. Taxonomic authorities for the botanical names will usually be found in the original references.

CHAPTER 2
Identification and Classification of Nutrients

2.1 Definition of Nutrients

Plants grow literally out of thin air, utilizing CO_2 as a primary source of carbon. Atmospheric oxygen is also essential, even though this reservoir has accumulated from the photosynthetic activities of plants over many millions of years. Oxygen is needed to support respiration in organs like roots, which possess no chloroplasts, or storage cotyledons that fail to develop functional stomata. In darkness, the whole plant needs an external source of O_2.

Plants also need water. Most plants take up water via their roots and lose it back to the atmosphere through stomata that are open by day (Milburn, 1979; Willmer, 1983). With insufficient water, stomata close or remain shut - preventing gas exchange and net CO_2 fixation by photosynthetic cells. Under natural conditions water supply is often the critical limiting factor for plant growth (e.g. Tunstall and Connor, 1975).

Besides these large-scale requirements for CO_2, O_2 and water, plants need a variety of mineral nutrients available from soil solution. For the early history of mineral nutrition, see Volume III of Steward's Treatise (Steward, 1963), and also Arnon (1958), Nicholas (1961), Epstein (1972) and Vose (1980). Although most macronutrients were recognized last century (Sachs, 1875), the number of nutrients regarded as essential has increased as a consequence of improvements in techniques for purification and analysis. Traces of a given chemical can now be removed more certainly from experimental solutions, and the residual concentrations can be accurately measured. Borate and silicate, for example, are readily leached from borosilicate glasses. Exceptional precautions are necessary in order to study the role of boron in plant nutrition, or to obtain media with minimal silica content (0.035 ppm) for control treatments (Campbell et al., 1975; Sangster, 1970 a,b; respectively).

To be classified as essential according to Arnon and Stout (1939), a nutrient element should conform to all of the following:

(1) normal growth or reproduction cannot occur without it
(2) it cannot be replaced with another nutrient
(3) the requirement should not be solely indirect, as would occur if the toxic effects of another substance were being alleviated.

A contemporary view of plants' nutritional requirements does not adhere rigidly to these three criteria. Authors such as Bidwell (1974, 1979) and Noggle and Fritz (1976) have pointed out that all three criteria cannot always be demonstrated for a given species under certain environmental circumstances, and that this should not automatically be construed as non-essentiality.

In practice, there has been a strong tendency to relax the second criterion concerning substitution. Nicholas (1961) introduced a broader concept, that of the "functional or metabolic" nutrient. This was meant to include "any mineral element that functions in plant metabolism irrespective of whether or not its action is specific". The concept of a "useful" nutrient has practically replaced that of an essential nutrient. "Useful" equals "capable of being utilized". The second classic criterion has misled generations of plant physiologists as to the way plants really operate. The flexibility of plants in compensating for nutritional inadequacies by improvising is an important property, often of great ecological significance (Crawford, 1976).

The mineral elements that can be regarded by functional criteria as being necessary for the growth of angiosperms are listed in Table 1. Perhaps Sr and Ba should also be considered (Lott, 1984). As there is no consensus on how to distinguish macronutrients from micronutrients, my inclusion of Fe with the micronutrients is arbitrary. Often Fe contents are found to be less than 50 ppm, but chemical analyses *per se* are no real guide to sufficiency. Tropical legumes such as *Phaseolus atropurpureus* and *Glycine wightii* can show foliar deficiency symptoms with Fe contents of 106 and 97 ppm respectively (Andrew and Pieters, 1972, 1976a).

Table 1.

Mineral Elements with Essential Functions in Angiosperms

Usually as macronutrients:	
	N, K, Ca, P, Mg, S, Si, Cl
Usually as micronutrients:	
	Fe, Na, Se, Mn, Zn, B, Cu, Mo, Co, Ni

2.2 Functions of Mineral Nutrients

2.2.1 Nitrogen

The two main sources of nitrogen for plants are nitrate, and ammonium, which may be either exogenous, or endogenous if N_2-fixing symbionts are contained within the root system. Nitrogen is essential as a component of amino and imino acids, amides, ureides, peptides, proteins, nucleic acids and their precursors, and secondary metabolites of many kinds (Harborne, 1982; Bell, 1984). The assimilation, translocation and storage of nitrogen are all complex processes, considered in more detail in subsequent chapters.

2.2.2 Potassium

Potassium is prominent as a transport electrolyte throughout the plant. It is the most abundant mineral cation in phloem sap (Section 4.2). The guard cells of stomata take up K^+ selectively as part of their mechanism for gaining turgor relative to adjacent epidermal cells (see Willmer, 1983). As a cofactor, K^+ may stimulate or facilitate many enzyme-catalysed reactions, some of which are discussed in Section 2.3.

2.2.3 Calcium

The principal fate of Ca^{2+} within the plant is incorporation into the primary cell walls as Ca pectate. Calcium is also the most abundant cation constituent of oxalate crystals present as vacuolar inclusions in leaf idioblasts (Lersten, 1983) and in protein bodies of vacuolar origin (Lott, 1984). There are strong indications that Ca^{2+} is implicated in the orientation of chloroplasts in response to light (Wagner and Klein, 1978) and in the geotropic response of whole roots (Hepler and Wayne, 1985). Ca^{2+} can act as a cofactor for enzyme-catalysed reactions, and as a regulator of energy metabolism (Section 2.3).

2.2.4 Phosphorus

Phosphate esters are indispensable metabolites, cofactors and precursors, e.g. of DNA and RNA. Phospholipids are essential components of membranes. Inorganic phosphate (Pi) itself is a key metabolite and acts as an effector or inhibitor of enzymes involved in intermediary metabolism. Most phosphorus in plants remains in soluble forms (Bieleski, 1973; Harper et al., 1982) until seeds develop, when phytate becomes the main insoluble reserve form of P in most species studied so far (Lott, 1984; Lott et al., 1984).

2.2.5 Magnesium

Magnesium is the central atom of chlorophyll and is almost always the most abundant divalent cation available as a cofactor for many enzyme-catalysed reactions (Section 2.3) particularly those involving phosphate-group transfer.

2.2.6 Sulphur

Sulphate itself serves an essential function as an anion, binding to specific sites on proteins and providing stability (e.g. phosphofructokinase, Kelly and Turner, 1969). Two of the polypeptide-forming amino acids contain sulphur: methionine and cysteine. The stability of many polypeptides depends upon disulphide bridges derived from the thiol groups of cysteine residues. Free thiol groups contribute to the active sites of some enzymes, and of some membrane proteins responsible for selective transport, e.g. of peptides into the scutellum of germinating cereal grains (see Dalling and Bhalla, 1984). Cofactors such as lipoic acid, Coenzyme A and glutathione contain functional thiol groups. A variety of secondary metabolites contain sulphur, e.g. glucosinolates and non-protein amino acids (Rosenthal, 1982; Bell, 1984). Some of these compounds may be multifunctional, fulfilling defensive and reserve functions in the seeds of some species.

2.2.7 Iron

Iron is the central atom in cytochromes, and is present in a number of 'non-haem' complexes with sulphur, including ferredoxin (Hall et al., 1975; Cammack et al., 1985). Very small quantities of ferredoxin are sufficient for normal chloroplast function: about 5 nmol per µmol of chlorophyll (Böhme, 1978). Plant metalloenzymes that contain Fe in the active site include nitrate reductase (Section 2.2.13), nitrite reductase, catalase, peroxidase and superoxide dismutase (del Río, 1983; Sevilla et al., 1984).

2.2.8 Manganese

Manganese has long been implicated in photosystem II as part of the O_2-evolving mechanism (Cheniae, 1970; Govindjee et al., 1978). A peroxisomal form of superoxide dismutase contains Mn in the active site (Sevilla et al., 1982), and Mn^{2+} may have a structural role in proteins such as the lectin concanavalin A (Becker et al., 1976). Mn^{2+} participates as a cofactor in many enzyme-catalysed reactions, e.g. acetolactate synthase (Davies, 1964), malic enzyme (Dilley, 1966) and PEP carboxykinase (Hatch et al., 1975). The role of cofactor is discussed in Section 2.3.

2.2.9 Zinc

Zinc is involved in the active site of cuprozinc forms of superoxide dismutase (Lastra et al., 1982), and carbonic anhydrase (Everson, 1970). The evidence linking carbonic anhydrase to photosynthetic CO_2 fixation is equivocal; Zn-deficient spinach leaves with low carbonic anhydrase activity are still capable of reasonable rates of CO_2 fixation (Randall and Bouma, 1973). Zinc may also be involved in a number of metalloenzymes that have been thoroughly studied in mammals and unicellular organisms, such as RNA polymerase, alcohol dehydrogenase, aminopeptidases and α-mannosidase (Hughes, 1981), but to date the evidence that corresponding enzymes from plants depend on Zn is largely circumstantial.

2.2.10 Boron

The precise functions of boron remain to be determined. The effects of boron deficiency include the malformation of leaves, death of the meristematic zones in shoot and root apices, and peculiarities in the differentiation of vascular tissues, in both monocotyledons and dicotyledons (Pissarek, 1980). These symptoms of deficiency are consistent with the idea that boron is necessary for cell division. Boron is also necessary for germination of pollen grains (Portnoi and Horovitz, 1977) and pollen tube extension growth. Many other ideas remain speculative; for discussion, see Nicholas (1961) and Clarkson and Hanson (1980).

2.2.11 Copper

The chloroplast pigment plastocyanin contains one essential Cu atom per polypeptide (Ramshaw et al., 1973). Phytochrome also contains copper. Some metalloenzymes contain Cu in the active site, e.g. polyphenoloxidase (Arnon, 1949), diamine oxidase (Hughes, 1981) and cuprozinc forms of superoxide dismutase present in the chloroplast (Lastra et al., 1982). Other enzymes may utilize Cu^{2+} as a cofactor (Section 2.3).

2.2.12 Selenium

A group of specialist plants that prosper on Se-rich soils has been identified (Shrift, 1969; Rosenthal, 1982); many are members of the genus *Astragalus*. These plants possess the ability to discriminate against selenomethionine and selenocysteine in the processes of transfer-RNA activation and aminoacylation that lead to the incorporation of cysteine and methionine into polypeptides. In plants for which excessive Se is toxic, these selenium analogues significantly replace sulphur-amino acids in polypeptides. The replacement of cysteine with selenocysteine is the more serious substitution, because normal disulphide bonds within or between polypeptides are fairly rigid, whereas such bridges with one or both S atoms replaced with Se allow more rotation; polypeptides so affected may not fulfil their normal functions as effectively. In susceptible plants, low concentrations of Se can be safely immobilized in polypeptides, and such plants can be extremely useful sources of Se for stock animals (Rosenthal, 1982).

In plants that tolerate high concentrations of Se, Se is

is incorporated into various non-protein amino acids, e.g. selenocystathionine. These compounds may accumulate in seeds (Dunnil and Fowden, 1967; Bell, 1984), even though the proteins stored in protein bodies in the same seeds may be devoid of Se (Lott and Volmer, 1979).

2.2.13 Molybdenum

Molybdenum was identified as an essential constituent of nitrate reductase from plant sources by Nicholas and Nason (1954). Mo-deficient plants accumulate nitrate, but when molybdate is supplied a rapid decline in nitrate content takes place, accompanied by a transient increase in nitrite concentration, and marked increases in amino N, amide N and protein contents (Spencer and Wood, 1954).

Other plant enzymes are now known to contain Mo, including xanthine dehydrogenase (Nguyen, 1986) and aldehyde oxidase(s) (Coughlan, 1980). Since the enzyme nitrogenase also requires Mo, plants such as legumes, which depend on the nitrogen-fixing activity of a prokaryotic symbiont are especially vulnerable to Mo deficiency. Apparently all the Mo-dependent enzymes also contain Fe and S in the active site (Coughlan, 1980).

2.2.14 Cobalt

As an essential constituent of the nitrogenase complex, Co is fundamentally important for the normal growth of plants that host N_2-fixing organisms in their roots. However, it has been difficult to identify any specific functions for Co^{2+} in plants generally, because of a lack of visible symptoms of deficiency (Nason and McElroy, 1963). The Co contents of plant tissues are usually so low that human dietary needs for cobalt cannot be sustained by a diet of plants alone; the inclusion of mushrooms or dairy products in vegetarian diets is therefore strongly recommended. Evidently herbivores gain Co in adequate amounts by virtue of consuming vast quantities of plant material.

Enzymes that depend upon adenosylcobamides or methylcobamides in mammals and micro-organisms are independent of cobalt in plants, and apparently have different cofactor requirements (Babior, 1975). Nevertheless, some enzymes from plants can utilize Co^{2+} as a cofactor *in vitro*. The enzyme thiamin pyrophosphate transferase, which catalyses the transfer of pyrophosphate from a nucleotide donor such as ATP to thiamin, requires a divalent cation as cofactor. Between 12 and 20 mM, Co^{2+} and Mg^{2+} are equally effective as most preferred cation; above this concentration range, Mg^{2+} is most effective, and below it, Mn^{2+} is most effective (Molin and Fites, 1980). Malic enzyme from apple fruit can function with Co^{2+} at a much faster rate than with Mg^{2+}, although Mn^{2+} is the most effective divalent cation (Dilley, 1966). In the chloroplasts of *Phaseolus vulgaris* leaves, there are two isoenzymes of phosphoglycollate phosphatase (Baldy et al., 1977; Verin-Vergeau et al., 1979). *In vitro*, one form prefers Co^{2+} to Mg^{2+}, whereas the other has the opposite preference. Since Mg^{2+} is the most abundant divalent cation in chloroplasts (Heldt, 1979), both forms of phosphoglycollate phosphatase are likely to function with Mg^{2+} *in vivo*.

There is no doubt that Co^{2+} can be utilized by certain plant enzymes if it is available, and in this sense, Co^{2+} is a useful nutrient. Whether there are any plant enzymes absolutely dependent on Co^{2+} remains to be determined.

2.2.15 Silicon

The need for exacting experimental precautions in studies on Si is emphasized by Harper et al. (1982), who concluded that quartz, apatite and whitlockite are all forms of silica that do not crystallize naturally in plants. However, the deposition of opaline silica ($SiO_2.nH_2O$) as phytoliths in leaves and other above-ground parts of plants has been thoroughly documented by Sangster (1970a,b). In many grass species, silica deposition occurs in an ordered sequence of cell types, beginning in idioblasts in expanding basal regions of each leaf. When expansion growth of each leaf is complete, deposition of phytoliths begins at epidermal long cells at the tip of each blade. Mature leaves develop extensive extracellular deposits of silica in the mesophyll, as well as in stomata, and in bulliform and long cells of the adaxial epidermis. These deposits become most extensive as the leaf senesces.

The movement of silicate in plants is mainly apoplastic (Aston and Jones, 1976). The importance of the structural contribution of silica to the leaves of grasses has been demonstrated by growing plants in a silica-minimal medium, when the leaves become more pliable, and lack their usual bristly texture (Sangster, 1970a,b). The deposition of silica in leaves could certainly "reduce transpiration and improve resistance to pathogens" (Bidwell, 1979), and by providing sharp edges to leaf blades, is certainly of adaptive value in deflecting the interest of some herbivores though not all.

Precise deposition of silica in endodermal cells in the roots of *Sorghum* (Sangster and Wynn Parry, 1976), in the inflorescence bracts of *Phalaris* (Sangster et al., 1983), and in pectin from citrus fruits (Schwartz, 1973) has also been documented. The general reluctance of plant physiologists to accept the structural role of silica is surprising, since the requirement for Si conforms to all three of the classic criteria for essentiality.

2.2.16 Chloride

Chloride is usually the most abundant anion available from soil solution, and is at very least useful as a counterion for essential cations, e.g. in xylem sap (Van Beusichem, 1983). The involvement of Cl^- in photosystem II has long been recognized (Bové et al., 1963), although specificity for Cl^- in this capacity has been questioned (Stout et al. 1954; Clarkson and Hanson, 1980). In numerous enzyme-catalysed reactions Cl^- assists by virtue of being an anion. More specific metabolic roles can be demonstrated, as in the activation of hydroxypyruvate reductase (Murray et al. 1973) and glutamate decarboxylase (Fonda, 1972).

Substantial requirements for Cl^- have been found for the growth of halophytes. Growth of *Sueda maritima* is maximally stimulated by external NaCl concentrations in the range 170-340 mM (Flowers, 1972). *S. australis* prefers a lower range (50-150 mM; Robinson and Downton, 1985).

In their studies of *S. australis*, Robinson and Downton have shown that the chloroplast concentrations of Na^+ (70 -140 mM) and Cl^- (80-120 mM) are held to these limits even under conditions of increasing external salinity that lead to more than 70-fold increases in the concentrations of both ions within the leaf. Thus halophytes and glycophytes appear to be similar in their requirements for Cl^- in the light reactions of photosynthesis, and in their abilities to regulate the concentrations of both Na^+ and Cl^- within the chloroplast.

2.2.17 Sodium

Over the past 30 years, Peter Brownell and his colleagues have provided substantial evidence that Na^+ is an essential nutrient. A series of treatments that will effectively minimize the Na^+ content of experimental media is fully described by Brownell and Wood (1957) and Brownell (1965). It is important to purify the air as well as the water and the individual mineral salts dissolved in the experimental media. In these original studies of *Atriplex vesicaria* (an Australian salt-bush), it was shown clearly that Na^+ is required for normal growth. Controlled additions of Na^+ alleviated deficiency symptoms, and when Na^+-insufficient plants were placed in medium containing 0.6 meq Na^+ per L, leaf contents of Na^+ increased more than 100-fold to about 80 μeq per g dry matter.

None of K^+, Rb^+ or Li^+ could substitute for Na^+. Given the composition of these media, and the stringent precautions taken in their preparation, it is clear that Na^+ is required *per se*.

Brownell and Crossland (1972) found that the most clear-cut requirement for Na^+ can be demonstrated in a variety of species with the C_4 mode of photosynthetic carbon assimilation (Section 3.1.5). Furthermore, CO_2 concentration can modulate the extent of deficiency symptoms in C_4 species: high CO_2 (1,500 μL per L) alleviates symptoms, but low CO_2 (15 μL per L) enhances them (Johnston et al., 1984).

Although the conclusion of these studies is that Na^+ is required as a micronutrient only for C_4 or some Crassulacean Acid Metabolism (CAM) species, perhaps a broader view of Na requirements might still be taken. Irrespective of photosynthetic mode, Australian species of *Atriplex* store NaCl in the bracts that surround the seeds (Beadle, 1952). The storage of salt in these bracts is of adaptive value, since the shed seeds will germinate only when sufficient rain has fallen to reduce the salt content of the bracts below concentrations that completely inhibit germination. The salt content of the bracts is about 0.9 M in water-saturated tissues. The extent of rainfall necessary for effective dilution, about 1 to 3 inches, is usually sufficient for the saltbush seedlings to complete a growth cycle, or in the case of annual species, their entire life-cycle (Beadle, 1952).

2.2.18 Nickel

The kind of effect envisaged in framing the third criterion of essentiality (Section 2.1) is well illustrated by the apparent requirement for Ni^{2+} described by Maranville (1970). He observed that the nitrate reductase activities of leaf extracts prepared from two C_4 species (sorghum, *Sorghum bicolor*, and sudan grass, *Sorghum sudanense*) can be considerably enhanced by the addition of 4 mM Ni^{2+} to the extraction medium. As cyanide is known to be generated in crude leaf extracts from cyanogenic glucosides, Maranville (1970) tested the hypothesis that the stimulatory effect of Ni^{2+} is due to its interaction with CN^-, which might otherwise inhibit nitrate reductase activity in the crude extracts. Assays of nitrate reductase activity in wheat leaf extracts, prepared with and without added Ni^{2+}, and to which known amounts of CN^- had been added, gave results consistent with the view that the effects of Ni^{2+} in *Sorghum* leaf extracts were entirely indirect; thus Ni is not considered to be one of the metals participating in the active site of nitrate reductase.

However, a specific enzymic function for Ni^{2+} has been determined - as an essential component of urease (Dixon et al., 1975, 1980a,b). This dependence on Ni is the reason for selective inhibition of seed forms of urease by hydroxamates or hydroxyurea (Dixon et al., 1980b; Kerr et al. 1983). Most of the seed forms of urease from jack bean (*Canavalia ensiformis*) comprise three or six polypeptide subunits (each of MW about 96,000), with two Ni^{2+} bound per subunit.

Two possible intracellular locations for urease in seeds are the protein body and the mitochondrion (Murray and Knox, 1977). During the development of soybean seeds (*Glycine max*), the measured urease activity of the cotyledons lags distinctly behind the accumulation of the enzyme protein, detected and estimated immunochemically (Fig. 1). Polacco and Sparks (1982) concluded that the formation of active urease molecules involves the insertion of Ni^{2+} at the active sites of preformed apoenzyme polypeptides. This interpretation is supported by the subsequent experiments of Winkler et al. (1983). When soybean plants of a normal

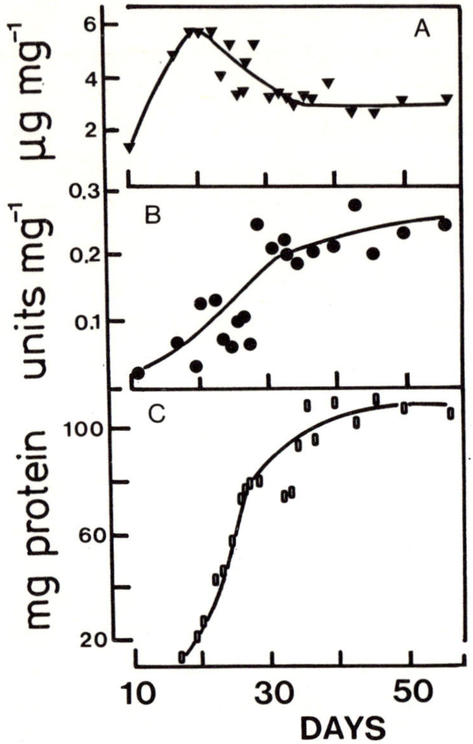

FIG. 1. Changes in the amount and activity of urease in developing soybean seeds: A, μg of urease protein per mg total protein; B, urease activity as units per mg protein; C, mg protein per seed. Data of Polacco and Sparks (1982).

* * * * * *

type were grown on Ni-minimal medium, the seeds that they produced exhibited <0.005% of the urease activity of Ni-sufficient seeds. However, the Ni-deficient seeds had a normal quantity of apoenzyme protein (Table 2). Thus the synthesis of urease polypeptides and their assembly into polymeric forms takes place independently of Ni^{2+} supply during seed development. Reconstitution of extracted apoenzyme and Ni^{2+} to form active urease does not take place *in vitro*. Nevertheless, when Ni-depleted seeds are allowed to imbibe Ni^{2+} from 2 mM solution, some incorporation of

Table 2.
Ureolytic Activity and Ni Content of Soybean Seeds.

Cultivar	Urease genotype	Ni content[a]	Ureolytic activity[b]	Urease protein[c]
Prize	normal	5.7	747	5.2
Itachi	null	3.7	0.08	0
Maple Presto normal:				
no added Ni		<0.01	0.013	4.1
Ni at 1 μg per L		0.06	7.2	4.3
Ni at 10 μg per L		0.81	284	5.2

[a] values are $\mu g\ g^{-1}$, determined by atomic absorption
[b] nmol urea hydrolysed $min^{-1}\ mg^{-1}$ protein
[c] μg antigen protein per mg total protein
Data of Winkler, Polacco, Eskew and Welch (1983).

* * * * * *

Ni^{2+} by the apoenzyme takes place, as evidenced by an increasing urease activity, even in the presence of cycloheximide to prevent new protein synthesis at cytoplasmic ribosomes (Winkler et al., 1983).

Some authors have suggested that because the enzyme urease has a restricted distribution among plant species, the associated Ni^{2+} requirement is of only limited importance. In legumes at least, the enzyme is widely distributed. In storage cotyledons, synthesis of the enzyme protein may be confined to the period of seed development, as observed in soybean (Fig. 1), and cowpea, *Vigna unguiculata* (Peoples et al., 1985a). Alternatively, urease synthesis may take place during development, then resume following imbibition and germination, as in the pea, *Pisum sativum* (de Ruiter and Kollöffel, 1983).

As to the function of urease, Polacco and Sparks stated that "a physiological role for urease has yet to be elucidated". There is some doubt now whether urease in the cotyledons of jack bean seedlings contributes to the mobilization of nitrogen from the guanidinoxy group of the non-protein amino acid L-canavanine, since this analogue of arginine is transported selectively from the cotyledons to the shoots of new seedlings with no detectable net loss (Rosenthal and Rhodes, 1984). However, in the same study, these authors demonstrated that urease could function *in vivo* to rapidly degrade either [^{14}C]urea or [^{15}N]urea injected into the cotyledons of 9-day old seedlings. Within 1.5 h of injection, only 18% of the label administered as [^{14}C]urea could be recovered from the seedlings (14.2% of this in the cotyledons), and only 4.4% of the label was in urea itself.

A common function for urease supported by all the available evidence is simply the mobilization of arginine, as the free amino acid (Young and Mason, 1972; de Ruiter and Kollöffel, 1983), and from the breakdown of proteins in the cotyledons. Arginine is not itself transported out of legume cotyledons; in pea, for example, the main exported forms of nitrogen are homoserine, asparagine and glutamine (Melcher, 1985).

A distinct requirement for Ni has also been demonstrated in barley, *Hordeum vulgare* (Brown et al., 1987). Plants progressively depleted of Ni over two preceding generations were supplied with zero, 0.6 or 1.0 μM $NiSO_4$. Grain yields and grain mass were not significantly affected by the absence of adequate Ni, but subsequent germination and seedling vigour were depressed at Ni contents less than a critical value of about 90 ng g^{-1} grain dry matter.

In contrast to the effects of exogenous Ni^{2+} on Ni-depleted soybean seeds, the depression in germination rate of Ni-depleted barley grains could not be remedied by allowing grains to imbibe for 24 h in $NiSO_4$ solution. Thus the requirement for Ni in cereals is associated with the process of seed development, and Ni is necessary for the life -cycle to be completed.

2.3 Specificity versus Adaptability of Cation Requirements

The ability of plants to substitute one nutrient for another contributes to their capacity to cope with variable nutrient supplies. Recognizing the possibilities and limitations of cation substitution will be helpful in relating the properties of isolated enzymes *in vitro* to the performance of organs and tissues in the living plant.

2.3.1 Monovalent Cations

One of the earliest established requirements for K^+ was that demonstrated for the ADP-glucose dependent starch synthase associated with starch granules isolated from the seeds of several cereals and legumes (Nitsos and Evans, 1969). Figure 2 shows that K^+ is more effective than NH_4^+ or Na^+ at equal concentrations. Furthermore, *in vivo*, seed tissues like developing endosperm are seldom likely to exhibit intracellular concentrations of NH_4^+ or Na^+ anywhere near those of K^+. In this case, the enzyme is unlikely to function effectively without K^+.

Asparaginase shows a strong preference for K^+ (Table 3). Although NH_4^+ is generated as a product of the reaction, by virtue of the strong specificity of the enzyme this NH_4^+ would fail to contribute effectively to a reduction in the need for K^+.

By comparison with these specific requirements for K^+, the NADH cytochrome *c* reductase associated with membranes of protein bodies can be stimulated almost equally well by K^+ and Na^+ (Plant and Moore, 1983). At equal concentration NH_4^+ was two- to three-fold more effective than K^+ or Na^+. Although the function of this enzyme is unknown, it seems that *in vivo* its univalent cation requirement can be met by any of these three cations according to local availability.

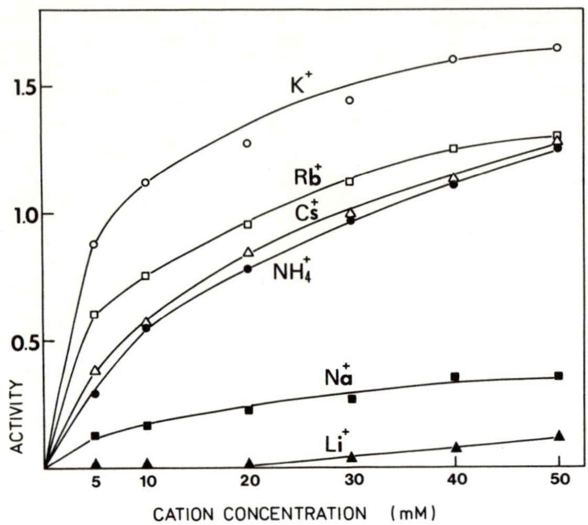

FIG. 2. The effects of varying univalent cation identity and concentration on starch synthase activity from sweet corn. Activity is mμmol ADP formed h^{-1} μg^{-1} protein. Data of Nitsos and Evans (1969).

* * * * * *

Table 3.

The Effects of Varying the Identity of the Univalent Cation on Asparaginase Activity from Developing Pea Seeds.

Cation[a]	Seedcoats	Cotyledons
Potassium	100.0[b]	100.0[b]
Sodium	38.8	26.6
Rubidium	36.3	15.5
Ammonium	7.5	1.1
Lithium	6.3	1.6
Cesium	5.9	0
Tris	5.6	1.1

[a] each cation was supplied at 33 mM as the chloride.

[b] activities are expressed as % of the maximum from each source. Data of Sodek, Lea and Miflin (1980).

2.3.2 Divalent Cations

The activity of a given enzyme must often depend upon a complex interplay among the various cations in solution in the same intracellular compartment. The relationships between Mg^{2+} and Mn^{2+} are particularly interesting in this regard; sometimes substitution is possible, sometimes not. Of the enzymes already mentioned (Sections 2.2.8; 2.2.14), malic enzyme (Dilley, 1966) and PEP carboxykinase show a strong preference for Mn^{2+}. However, the two forms of acetolactate synthase present in developing pea seeds provide an example where substitution is possible (Davies, 1964). The reaction catalysed by acetolactate synthase is a common step in the syntheses of branched-chain amino acids, and is specifically inhibited by sulfonylurea herbicides (Ray, 1984). One isoenzyme has a pH optimum at 6.0, a high Km for pyruvate (43 mM) and prefers Mn^{2+}; the other has a pH optimum at 8.5, a higher affinity for pyruvate (Km = 3.2 mM) and prefers Mg^{2+} (Table 4). The cation preferences are not strict; each isoenzyme can operate by accepting the alternative cation, and for both forms, Mn^{2+} is effective at a lower concentration than Mg^{2+}.

Table 4.
Effects of Mg^{2+} or Mn^{2+} on the Activities of Two Forms of Acetolactate Synthase from Developing Pea Seeds.

Cofactors added:	Acetolactate produced[a] pH 6.0	pH 8.5
None	0.06	0.18
Mg^{2+} (5 μmol)	0.10	0.26
Thiamin pyrophosphate	0.19	0.39
TPP + Mg^{2+} (5 μmol)	0.91	1.04
TPP + Mn^{2+} (0.5 μmol)	1.17	0.94

[a] μmol acetolactate formed h^{-1} per mg protein
Data of Davies (1964).

Glutamine synthetase catalyses the ATP-dependent formation of glutamine from NH_4^+ and glutamate, and provides the principal mechanism for the assimilation or reassimilation of NH_4^+ in plant cells. This reaction depends upon Mg^{2+} as cofactor, as would be expected for a reaction involving ATP. It has recently been suggested that a second Mg^{2+} may be required to attach ammonia in the correct position within the active site, forming a bridge between NH_3 and an essential thiol group (Evstigneeva et al., 1985). However, since the pH optima for the glutamine synthetase reaction are well below the pK' for NH_4^+ (ca. pH 9), the reactive species is probably NH_4^+, not NH_3. The involvement of a second Mg^{2+} therefore remains speculative, but the involvement of one Mg^{2+} with the negatively charged phosphate of the ATP is essential, and highly specific.

The chloroplast form of glutamine synthetase from pea leaves acting with Mg^{2+} is inhibited by Mn^{2+}, Ca^{2+} and Zn^{2+} (but not Co^{2+}). The inhibition by alternative cations is only partly due to a distinct shift in pH optimum induced by each cation (O'Neal and Joy, 1974). Glutamine synthetase activity from lupin nodules is far more severely inhibited by either Mn^{2+} or Ca^{2+} than the pea chloroplast isoenzyme, with I_{50} values of 0.025 and 0.189 mM respectively (Chen and Kennedy, 1985). Another difference compared to the pea leaf (chloroplast) isoenzyme is its complex inhibition by Co^{2+}, which is partially relieved by glutamate.

Apparently Mg^{2+} and Mn^{2+} confer distinct properties on the glutamine synthetase enzyme molecules associated with each cation. Despite being an effective inhibitor of Mg^{2+}-dependent glutamine synthetase activity, Mn^{2+} is the preferred cation for the γ-glutamyl exchange reaction catalysed by the same enzyme. The assay for the exchange reaction is usually performed by measuring the formation of γ-glutamyl hydroxamate from glutamine and hydroxylamine in the presence of arsenate, ADP and Mn^{2+} (Shapiro and Stadtman, 1970). The rate of the exchange reaction is often 8- to 10-fold higher than the glutamine synthetase reaction conducted with the same amounts of enzyme protein (O'Neal and Joy, 1974; Harel et al., 1977; Murray and Kennedy, 1980). Further evidence that glutamine synthetase may take up different functional conformations in the presence of alternative cations is seen in the observation that attempts to purify the enzyme from lupin nodules by affinity chromatography on 2',5'-ADP-Sepharose succeed in the presence of Mn^{2+}, but no association of glutamine synthetase and this medium occurs in the presence of Mg^{2+} (Chen and Kennedy, 1985).

2.3.3 Cation Substitution and the Assessment of Nutrient Deficiencies

Assessment of the nutrient status of growing plants by the analysis of leaves or other plant parts is time-consuming, and not always reliable. The problem has been discussed by Besford (1978), Smith (1978) and del Río (1983). The possibility that isoenzyme analysis might be faster for diagnostic purposes has been emphasized by del Río (1983), who points out the advantages of assaying the superoxide dismutase isoenzymes. This enzyme system is believed to protect cells from the potentially damaging effects of the superoxide radical, O_2^- (Halliwell, 1974; Fridovich, 1983).

The best known forms are cuprozinc metalloenzymes (see Hughes, 1981) and in plants, this type is found in the chloroplast (Lastra et al., 1982; Duke and Salin, 1983). The Fe-containing form is rare in angiosperms, but when present, as in the lemon, is found in the chloroplast (Salin and Bridges, 1982) and possibly other organelles as well (Sevilla et al., 1984). A third type, containing Mn, has been localized in peroxisomes (Sevilla et al., 1982; del Río et al., 1983).

A corollary of the existence of multiple forms of an enzyme with distinct metal requirements is that deficiencies in certain cations might lead to declines in the corresponding form(s) of the enzyme, coupled perhaps with compensating increases in the amounts or activities of the alternative form(s). For superoxide dismutase, this can be tested by test-tube assays, and by detecting bands of enzyme activity in gels following electrophoresis of crude extracts. To identify the functional metals associated with each band of activity, selective inhibition by CN^- (1 mM) and H_2O_2 can be scored. The Cu.Zn type is sensitive to both inhibitors; the Fe type is sensitive to peroxide but insensitive to CN^-, and the Mn type is insensitive to both.

Table 5 records data for the leaves of lemon trees grown under full-nutrient, Fe-deficient and Mn-deficient conditions. In response to either deficiency, the activities of all the superoxide dismutase isoenzymes decrease compared to control plants. The Fe-deficient leaves show activities close to the control, whereas the Mn-deficient leaves show a lower total activity, and also a much reduced Zn content (Table 5). In pea, a species lacking Fe-dependent forms, the effect of Mn deficiency is to selectively lower the activity of the Mn-dependent isoenzyme in the leaf, while increasing the contribution of the cuprozinc forms to the total activity in the leaf (del Rio et al., 1978).

Copper deficiency has no pronounced effect on total superoxide dismutase activities of leaves of several ages, but considerably depresses the activities of ascorbate oxidase, diamine oxidase and cytochrome oxidase in young leaves and shoots of subterranean clover (Walker and Loneragan, 1981). Zinc deficiency leads to marked increases in ribonuclease activity, correlated with lower leaf nitrogen contents (Johnson and Simons, 1979).

There appears to be considerable scope for exploring the

Table 5.
Superoxide Dismutase Activity and Metal Contents of Leaves from Lemon Trees Grown under Three Nutrient Regimes.

	Enzyme activity (units mL^{-1})			Leaf content[c]		
Growth regime	Total	Fe+Mn[a]	Cu.Zn[b]	Fe	Mn	Zn
Full-nutrient	71.65	30.95	40.70	103	81	77
Fe-deficient	63.08	27.55	35.53	46	84	78
Mn-deficient	47.68	20.44	27.24	67	10	23

[a] total cyanide-insensitive component
[b] cyanide-sensitive component
[c] each as $\mu g\ g^{-1}$ leaf dry matter
Data of Sevilla, del Río and Hellín (1984).

* * * * * *

diagnostic potential of enzyme and isoenzyme assay, provided that suitable organs or tissues are chosen for analysis, and cultivar differences are recognized and documented.

2.4 Immobility - An Unreliable Basis for Nutrient Classification

Mineral nutrients have traditionally been classified in a quantitative fashion, with an arbitrary division between macronutrients and micronutrients (Section 2.1). Another method of classification is to consider whether each nutrient can be withdrawn from the cells to which it was originally committed, and remobilized via phloem to some developing part(s) of the plant. Nutrients are said to be 'immobile' if they are not readily relocated via phloem under conditions of imposed external deficiency. Clearly this becomes an important factor in the allocation of nutrients from current uptake to developing fruits and seeds.

Silicon is immobilized irreversibly by deposition as phytoliths (Section 2.2.12). Other nutrients regarded as immobile are Ca, Fe, Zn, Cu, Mn, Mo and B. However, an assessment of this kind may well be an oversimplification.

First, there are very strong species differences. Boron, for example, may be immobile in the classroom tomato seedling, but mobile in fruiting peanuts (Campbell et al., 1975). A systematic nutritional survey conducted by the C.S.I.R.O. Division of Tropical Crops and Pastures (Brisbane, Australia) illustrates the importance of species when making assessments of mobility or immobility. In none of the five species I have chosen as examples were the symptoms of deficiency consistent with complete immobility of all the 'immobile' nutrients. Boron is mobile in *Phaseolus atropurpureus* (Andrew and Pieters, 1972), Ca is mobile in *Glycine wightii* (Andrew and Pieters, 1976b), Zn and Cu are mobile in *Chloris gayana* (F. W. Smith, 1973) and partially mobile in *Lotononis bainesii* (Andrew and Pieters, 1976a) and *Stylosanthes humilis* (Jones and Clay, 1976). For these five species, only Fe and Mn were consistently immobile. For the two legumes, a deficiency of Mo gave rise to symptoms consistent with immobility of Mo only when plants were supplied with nitrate as nitrogen source, rather than permitted to form N_2-fixing nodules, when the symptoms of Mo-deficiency were indistinguishable from those of nitrogen deficiency.

Are there circumstances when relatively immobile nutrients such as Fe or Mn can be mobilized? Nable and Loneragan (1984) examined this question in relation to the growth of sub clover. The first-established leaves retained all the ^{54}Mn they had derived from an earlier supply after plants had been transferred to a nutrient solution containing no Mn. These authors also considered that leaf-loss of Mn under field conditions probably reflected leaching rather than mobility in phloem. In contrast, Vose (1963, 1980) used ^{54}Mn to demonstrate that in oats, Mn applied to the leaves is redistributed preferentially to developing leaves or grains. Studies on both of these species indicate that some mobilization of Mn from roots can occur, according to the stage at which a deficiency is imposed.

Apart from between-species differences in the extent of mobility exhibited by individual nutrients, considerable variability also occurs within a single species or cultivar. Mobility may be affected by stage of development, or the current rate of supply of the nutrient under consideration (Loneragan et al., 1976). In view of this discussion it must be concluded that any attempt to classify nutrients on the basis of phloem mobility versus immobility is unworkable.

CHAPTER 3

Assimilation and Translocation of Carbon and Nitrogen

This chapter focusses on the functioning of leaves, stems and roots as supply organs. The assimilation of carbon is a consequence of photosynthetic CO_2 fixation in leaves or equivalent organs, whereas the primary assimilation of inorganic forms of nitrogen occurs mainly in leaves, roots and stems, in proportions that vary with source, stage of development, and plant species or variety.

3.1 Photoassimilation of Carbon Dioxide

Detailed accounts of progress in understanding photosynthesis and leaf development may be found in books and articles by Bradbeer (1977), Thimann (1987), Edwards and Walker (1983), Sesták (1985), Badger (1985) and Ting (1985). The most important thing to be said about photosynthesis at the outset is that there is no simple equation capable of portraying yields of reducing potential and ATP from the light reactions, and balancing these against the precise needs of the Benson-Calvin CO_2-fixation cycle. All equations purporting to show exact relationships between oxygen evolved and carbon dioxide fixed are erroneous. The reason that simple equations are inadequate is that photosynthesis is a dynamic and complex process, subject to a number of interacting controls, and there is no single product representing assimilated carbon, nitrogen, sulphur and phosphorus.

3.1.1 C_3 Species

In C_3 plants the primary fixation of CO_2 is catalysed by the enzyme ribulose-1,5-bisphosphate (RuBP) carboxylase, which is confined to chloroplasts or their plastid precursors. Two molecules of the C_3-acid 3-phosphoglycerate are formed per molecule of CO_2 fixed (Fig. 3). Triose phosphate is produced by reduction of 3-phosphoglycerate, a step requiring both NADPH and ATP, and catalysed by a chloroplast-specific form of triose phosphate dehydrogenase. Most of the triose phosphate produced takes part in a series of transformations leading to regeneration of RuBP, but some is eligible for export by a specific transporter in the inner envelope, which exchanges triose phosphate for imported inorganic phosphate (Flügge and Heldt, 1984).

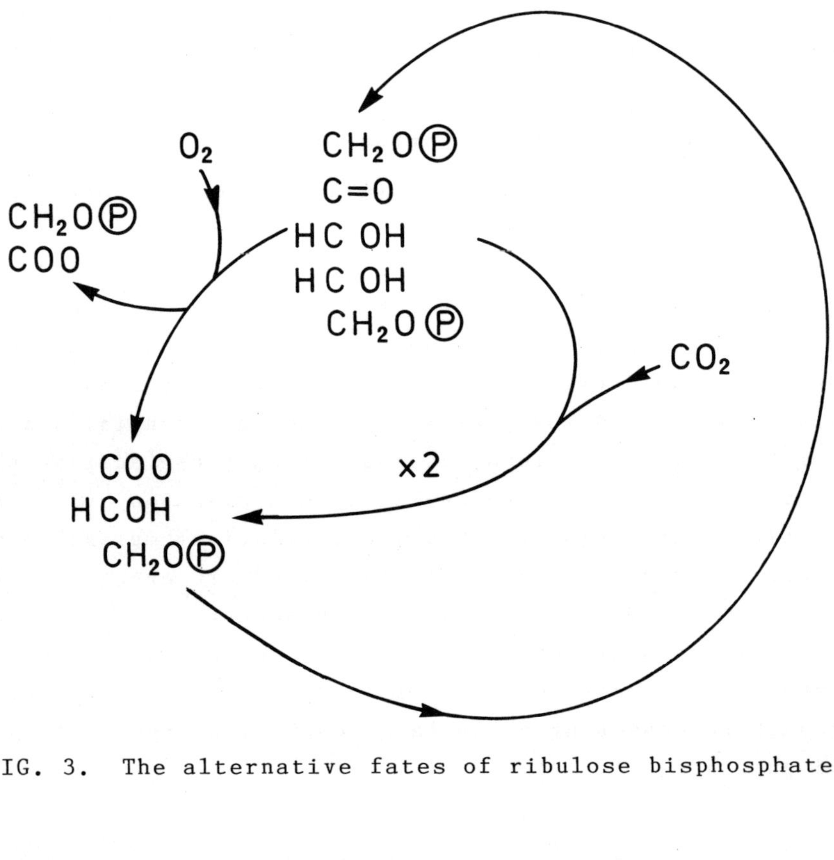

FIG. 3. The alternative fates of ribulose bisphosphate.

The rate at which triose phosphate and Pi exchange can inhibit or promote the overall rate of photosynthesis, because the chloroplast envelope restricts the transfer of many other phosphorylated metabolites. The direct export of ATP, NADPH and key sugar phosphate intermediates of the Benson-Calvin cycle is effectively prohibited, hence triose phosphate represents the major form in which reducing potential is transferred from the chloroplasts to the cytoplasm. If the availability of Pi is diminished in the cytosol, as would occur if the rate of conversion of triose phosphate to non-phosphorylated derivatives slowed, then lower internal concentrations of Pi in the chloroplasts will limit the possible rate of photosynthetic phosphorylation, and consequently CO_2 fixation. Regulatory possibilities are discussed further in Section 3.2.1.

The concept of a smooth-running carbon fixation cycle with an effective control over the diversion of intermediates to the cytoplasm is complicated by another reaction undertaken by RuBP carboxylase -- the oxygen-dependent cleavage of RuBP to one molecule each of 3-PGA and the C_2-compound, phosphoglycollate (Fig. 3). Non-enzymic oxidation of RuBP to yield these products has been known since the time the main pathway of CO_2 fixation was elucidated (Goodman et al., 1955). This reaction was suggested as a possible source of phosphoglycollate (Murray, 1969) when evidence had accumulated that phosphoglycollate is the principal immediate precursor of glycollate in the leaves of C_3 species (Richardson and Tolbert, 1961a,b; Kearney and Tolbert, 1962; Yu et al., 1964).

The demonstration that RuBP carboxylase also possesses RuBP oxygenase activity (Bowes et al., 1971; Lorimer et al., 1973) was a major breakthrough in confirming the origin of glycollate and explaining the well known Warburg effect - the inhibition of net CO_2 fixation in the leaves of C_3 species brought about by increased partial pressures of O_2. This inhibition is now seen as an inevitable consequence of competition between CO_2 and O_2 at catalytic sites of activated RuBP carboxylase molecules (Laing et al., 1974; Lorimer, 1981; Ogren, 1984; Portis et al., 1986).

3.1.2 Photorespiration and Glycollate Metabolism

That phosphoglycollate is derived from RuBP *in vivo* has been supported by studies with $^{18}O_2$ supplied to leaves of C_3 plants (*Spinacia oleracea*, *Helianthus annuus* and *Atriplex hastata*) at low partial pressures of CO_2 (Berry et al., 1978). Glycollate, glycine and serine were labelled rapidly after brief exposure of the leaves to $^{18}O_2$, and only one of the oxygen atoms of the carboxyl group of each compound received the label. After only 5 to 10 seconds, the labelled O atom in glycollate attained an isotopic enrichment that was 50 to 70% of that in the $^{18}O_2$ supplied; after 40 and 180 seconds respectively, glycine and serine exhibited the same degree of enrichment. These data not only confirm the derivation of phosphoglycollate and hence glycollate from RuBP, but also demonstrate the expected formation of serine from glycine derived from glycollate (Fig. 4).

This pathway of glycollate metabolism is the major contributor to 'photorespiration', which consists of all pathways or reactions consuming O_2 or generating CO_2 in the light. Oxygen is consumed in the cleavage of RuBP to form phosphoglycollate, and in the conversion of glycollate to glyoxylate (Figs 3, 4); CO_2 is evolved from the carboxyl group of glycine degraded at appropriate rates in leaf mitochondria (Bird et al., 1972; Kumarasinghe et al., 1977; Keys, 1980). However, is there really sufficient evidence to regard the glycollate pathway as the Photorespiratory Carbon Oxidation Cycle (PCO cycle), as proposed by Berry et al. (1978)? Certainly these authors found that

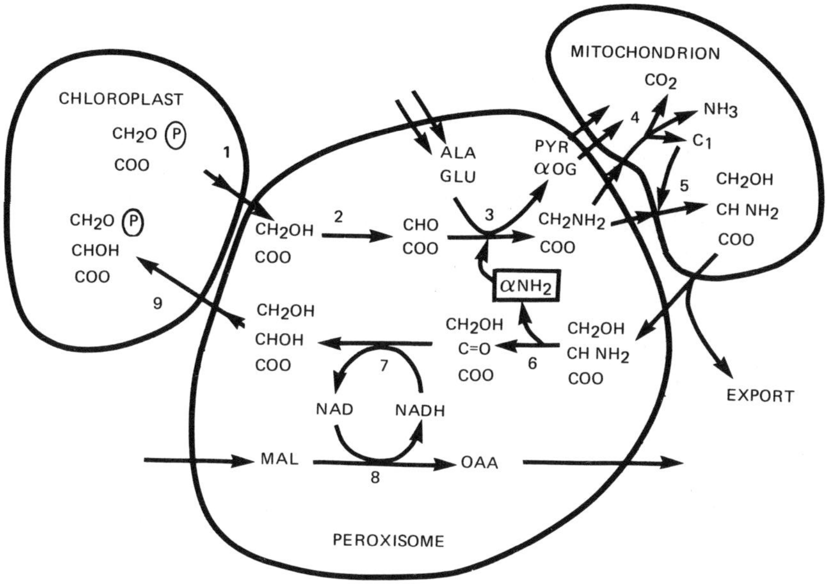

FIG. 4. The glycollate pathway. Enzyme key: 1, phosphoglycollate phosphatase; 2, glycollate oxidase; 3, glyoxylate aminotransferases; 4, glycine decarboxylase and C_1-unit interconversion; 5, serine transhydroxymethylase; 6, serine:glyoxylate aminotransferase; 7, hydroxypyruvate reductase; 8, malate dehydrogenase; 9, glycerate kinase.

* * * * * *

the carboxyl oxygen atoms of 3-PGA became labelled with ^{18}O in long-term feeding experiments (20 or 40 mins). But the enrichment of these carboxyl atoms was only 14 ± 4 % of that observed in serine. Berry et al. (1978) assumed that labelled serine was the sole precursor of 3-PGA, and explained the diminished labelling of 3-PGA as the result expected from a one-in-five dilution with unlabelled 3-PGA derived from RuBP carboxylase activity with unlabelled CO_2 in the chloroplasts. The short-comings of this argument are discussed further below.

The conversion of serine to 3-PGA depends on the ability of peroxisomes to take up serine following its release from mitochondria. Peroxisomes have long been recognized as the location of glycollate oxidase, catalase and amino-transferases donating amino groups to form glycine (Tolbert et al., 1968, 1970; Nakamura and Tolbert, 1983). The role of peroxisomes in transforming glycollate to glycine has been obvious, but the way in which they might take part in a return of C_3-units from serine to 3-PGA has been equivocal. On the basis of the early evidence, Tolbert (1973) suggested that glycerate could be generated from 3-PGA by phosphatase action in the chloroplasts, and then could be converted to hydroxypyruvate and serine in the peroxisomes, thus providing a second pathway for serine synthesis.

For serine to be converted to 3-PGA, the first enzymic step inside the peroxisome is removal of its amino group to form hydroxypyruvate (reaction 6 in Fig. 4). This step is catalysed by serine:glyoxylate aminotransferase. By comparison, enzyme-catalysed transfer of amino groups from serine to oxaloacetate or α-oxoglutarate is insignificant (Walton and Woolhouse, 1986), although donation to pyruvate is possible (Wallace et al., 1984). However, glycine formation is not totally dependent on the activity of the serine:glyoxylate aminotransferase. In their studies on the development of the fifth leaf of pea seedlings, Ta and Joy (1986) estimated that [^{15}N]serine supplied exogenously contributed only 38% of the amino groups received by glycine; [^{15}N]asparagine contributed 7% via the same aminotransferase, and jointly, [^{15}N]glutamate plus [^{15}N]alanine contributed more than half the amino nitrogen acquired by glycine. These estimates confirm earlier *in vitro* studies with isolated peroxisomes showing that glutamate and alanine preferentially contribute to the formation of glycine (Kisaki and Tolbert, 1969; Yu et al., 1984). Thus glycine synthesis can be maintained by glutamate and alanine if substantial proportions of the serine produced from glycine do not enter the peroxisomes (Section 3.1.3).

The second step in the sequence from serine to 3-PGA, reduction of hydroxypyruvate to glycerate, has long seemed plausible. The NADH-dependent glyoxylate reductase originally localized in peroxisomes by Tolbert et al. (1968, 1970) prefers hydroxypyruvate to glyoxylate as substrate. Hydroxypyruvate reductase activity increases in concert with phosphoglycollate phosphatase, glycollate oxidase, and catalase activities when dark-grown bean leaves are illuminated (Murray et al., 1973). *In vivo*, this activity could be regulated by local chloride ion concentration, and the availability of NADH. Replenishment of NADH consumed in the reduction of hydroxypyruvate to glycerate could be effected by the import of malate (Fig. 4), and the action of a peroxisomal form of malate dehydrogenase (Rocha and Ting, 1971).

The final step for conversion of serine to 3-PGA needs glycerate kinase activity and a supply of ATP. Two lines of evidence now firmly place this step inside the chloroplast, rather than in the cytosol. First, glycerate kinase has been localized in intact chloroplasts (Usuda and Edwards, 1980). Secondly, glycerate can be taken up by chloroplasts and metabolized (Heber et al., 1974). Howitz and McCarty (1986) have proposed that the same transporter in the inner envelope is responsible for both the uptake of glycerate and the export of glycollate. Since each of the glycerate molecules available for uptake would have been derived from two glycollate molecules, a precise one-for-one exchange cannot be the sole mechanism for glycollate export from the chloroplast. An exchange of pyruvate for glycollate seems more logical, since pyruvate is needed to support the considerable rates of fatty acid synthesis occurring inside the chloroplast (Williams and Randall, 1979). Perhaps the export of glycollate is not linked to the import of any single metabolite.

Overall, the evidence suggests that conversion of serine to 3-PGA is possible, but not necessarily stoichiometric as asserted by Berry et al. (1978). The small degree of ^{18}O enrichment of 3-PGA observed by these authors can be explained in other ways. Since their experiments were performed at the compensation point for CO_2, refixation of ^{18}O into 3-PGA should have occurred via the $C^{18}O_2$ derived from the carboxyl groups of the glycine molecules degraded in the course of serine synthesis in the mitochondria. Decarboxylation of glycine is the principal source of endogenous CO_2. The claim that no refixation of CO_2 from this source had occurred in periods as long as 40 mins at the CO_2 compensation point ran counter to what should have been expected on the basis of many previous studies of photorespiration (e.g. Ludwig and Canvin, 1971; Laing and Forde, 1971; Martin et al., 1972).

Refixation of $C^{18}O_2$ is a sufficient explanation for the labelling of 3-PGA in the long-term experiments of Berry et al. (1978); there is no evidence that any glycollate-derived serine was converted to 3-PGA under these conditions. The rush by most contemporary authors to adopt the PCO cycle terminology is totally unwarranted. In fact, the reactions of glycollate metabolism (Fig. 4) do not constitute a cycle in the conventional sense of the term. No acceptor molecule is regenerated, as in the Krebs cycle or the Benson-Calvin cycle. Furthermore, the conversion of serine to 3-PGA involves reduction, not oxidation. This sequence is better regarded as a salvage loop for C_3-units - possibly of limited significance in young leaves of plants that are not flowering. Serine does not accumulate in such leaves (Bauer et al., 1977a; Wallace et al., 1984) and some conversion of [2-^{14}C]glycine to sucrose can occur in the light (Miflin et al., 1966). The capacity for conversion of serine to 3-PGA is lost as leaves age, and hydroxypyruvate reductase activity declines (Gordon et al., 1978).

3.1.3 The Significance of Glycollate Metabolism for Serine Export

Frequently glycollate metabolism has been described as wasteful, and it has been claimed that C_3 plants might be more productive without photorespiration. How inefficient is this allegedly wasteful pathway? Each serine costs the release of one previously assimilated carbon as CO_2, as well as the phosphate-ester bond energy given up when phosphoglycollate is hydrolysed to glycollate and Pi (x 2 per serine). Another cost to be considered is the ATP and reducing potential needed to reassimilate the NH_4^+ released by the degradation of glycine, since this will take place mainly via glutamine synthetase (Section 2.3.2) and GOGAT (glutamine:oxoglutarate amidotransferase, or glutamate synthase) in the chloroplasts (Section 3.3.2).

One each of ATP and NADH will be required for each molecule of glycerate recovered from serine and reinstated in the chloroplasts as 3-PGA, but this will not often be the major fate for serine produced by this pathway. The extent to which some of these metabolic costs may be offset by reactions in the mitochondria is still uncertain. Even at their highest estimates these costs are not prohibitive, compared, for example, with the synthesis of polypeptides, involving the costs of each amino group plus one ATP for every amino acyl-tRNA formed, or with the synthesis of fatty acids from acetyl-CoA inside the chloroplasts, or with the assimilation of thiol groups from SO_4^{2-}, or amino groups from NO_3^- (Kennedy, 1986).

The costs of glycollate metabolism must be set against the benefits. Unlike the RuBP carboxylase reaction, the RuBP oxygenase reaction produces protons likely to be important for the acid-base balance of chloroplasts in C_3 plants, especially during nitrate assimilation (Chapter 6 of Kennedy, 1986; Section 3.3.1). Furthermore, glycollate metabolism is quantitatively important as a source of C_1-units associated with tetrahydrofolate, and glycine and serine *per se*. All are precursors of other essential compounds in leaves, including methionine (from C_1-units; Clandinin and Cossins, 1972, 1975), protoporphyrin rings for pigments (glycine), purine rings (glycine, and C_1-units derived from glycine; Lehninger, 1970; Lovatt, 1983), and tryptophan (serine). Glycine and serine are incorporated into newly synthesized polypeptides, together accounting for more than 40% of label present in protein following photoassimilation of $^{14}CO_2$ by pea leaves (Pate, 1966; Lewis and Pate, 1973).

Serine can be the predominant recipient of label among all amino acids and amides in leaves of C_3 species following photosynthesis with $^{14}CO_2$, e.g. in soybean (Nelson et al., 1961), sunflower (Atkins and Canvin, 1971) and cottonwood (*Populus deltoides*; Dickson and Larson, 1975). Two environmental factors which commonly increase the glycine and serine pool sizes in leaves and the extent of their labelling from $^{14}CO_2$ are water stress (Lawler, 1978; Goyal and Gupta, 1985) and low light intensity (see Wardlaw, 1982). In darkness, [^{14}C]glycine supplied to pea leaves is converted mainly to serine, with less than 1% of the total label in products entering sucrose (Miflin et al., 1966).

Serine is one of the major translocated nitrogenous solutes in the phloem sap of many hardwood tree species, and clearly originates in the leaves (Tromp and Ovaa, 1971). The extent to which serine is translocated from leaves depends on ontogenetic factors, such as stage of development (Nelson et al., 1961), as well as environmental factors. It is highly significant that the onset of flowering triggers an increase in the rate of photorespiration in all the leaves of a bean plant (Fig. 5). The rates of photorespiration in the leaves of chick pea plants are also maximal during flowering and early fruiting (Hooda et al., 1986). In pea plants, serine accumulates to become the most abundant nitrogenous solute in fully illuminated unstressed leaves subtending young pods (Murray, 1986a).

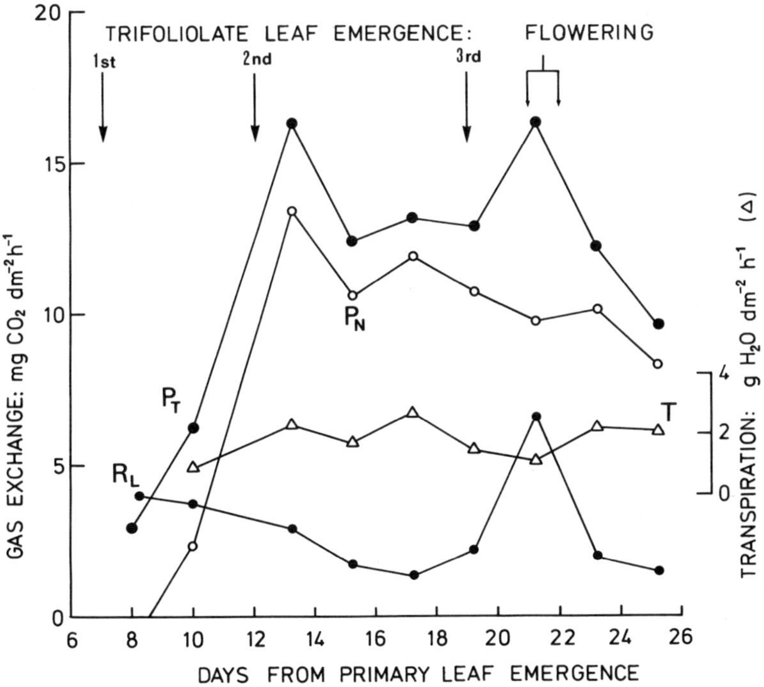

FIG. 5. Photosynthesis and transpiration of the first trifoliolate leaf of developing bean plants (*Phaseolus vulgaris* cv. Pencil Pod Black Wax). P_T, total photosynthetic CO_2 fixation; R_L, photorespiration; P_N, net photosynthetic CO_2 fixation. The times of leaf emergence and flowering are indicated by arrows. Data of Fraser and Bidwell (1974).

Labelling studies confirm that recently-formed serine is translocated from leaves, and is important for the nutrition of developing fruit-walls and seeds. This occurs in common thornapple (Lewis, 1975), pea (Lewis and Pate, 1973), broad bean (Kipps and Boulter, 1974), white lupin (Pate et al., 1974) and soybean (Fellows et al., 1978). Serine is often the fourth or fifth most abundant amino compound secreted by the seedcoats into the embryo sac (Murray, 1983b; Peoples et al., 1985a) and commonly serves as a precursor of glycine, methionine, cysteine and tryptophan, all of which are incorporated into the reserve proteins of the embryo.

3.1.4 The Importance of Glycollate Metabolism Assessed with Mutants

It is possible to determine what happens to C_3 plants when glycollate formation and metabolism are prevented. This question can be approached experimentally by varying the availability of O_2 or CO_2, or by studying mutants with impaired enzyme activities or transport functions (Somerville, 1984, 1986). Implementing the former approach, Powles et al. (1979) have shown that photorespiration can protect chloroplasts against photoinhibition. This is the damage caused to the photosystems when reducing potential cannot be fully utilized (Powles, 1984). Photoinhibition can be imposed on attached leaves by illuminating them in the absence of CO_2 and at low partial pressures of O_2. The effects of photoinhibition can be substantially prevented by admitting CO_2 at partial pressures corresponding to the compensation point, or O_2 at sufficient concentrations to promote endogenous CO_2 evolution via the glycollate pathway.

Somerville (1984, 1986) and his colleagues have studied several mutants of *Arabidopsis thaliana* (Cruciferae) with specific defects in individual enzymes of glycollate metabolism. Those affected are phosphoglycollate phosphatase, serine:glyoxylate aminotransferase, glycine decarboxylase and serine transhydroxymethylase. Under conditions of low O_2 availability and normal CO_2 supply (350 ppm), photosynthetic carbon metabolism is similar, both quantitatively and qualitatively, in each mutant and the wild type. But under normal conditions (21% v/v O_2), photosynthesis is rapidly inhibited in all the mutants. After several days under standard conditions, the mutants show symptoms of severe stress. These symptoms can be reversed by periods of darkness, or by diminishing the O_2 content of the atmosphere. This genetic approach to blocking the activity of an individual enzyme *in vivo* is far more specific and certain than the provision of supposed enzyme inhibitors, since it is always possible for an applied compound to have multiple effects. For example, α-hydroxysulphonates do not act solely as competitive inhibitors of glycollate oxidase (Murray and Bradbeer, 1971).

There is sufficient evidence to indicate that if the main pathway of photorespiration were eliminated from C_3 species, the resulting plants would be non-viable. If the synthesis and metabolism of glycollate were truly wasteful and served no useful functions, it is improbable that so many C_3 species would have survived for so long.

3.1.5 C_4 Species

The initial carboxylation reaction of C_4 species and CAM species is catalysed by phosphoenolpyruvate (PEP) carboxylase, yielding oxaloacetate. Short-term feeding studies with $^{14}CO_2$ show clearly that malate and aspartate receive most of the label accumulated initially (Kortschak et al., 1965; Hatch and Slack, 1966), a distribution explained by the action of malate dehydrogenase and aspartate aminotransferase activities respectively. Malate or aspartate or both are regarded as the stable derivatives of CO_2 fixation in the ground mesophyll cells. From these cells they are transported to bundle sheath cells, a typical feature of leaves with this mechanism of photosynthesis ('Kranz' anatomy).

Directly or indirectly, the fate of the C_4-acids transferred to bundle sheath cells is decarboxylation, and the generation of C_3-units returned to the mesophyll cells. Far from being considered 'wasteful', this immediate decarboxylation is perceived as a mechanism for concentrating CO_2 in the chloroplasts of the bundle sheath cells, where RuBP carboxylase may function with less effective competition from O_2, and so less diversion of newly fixed carbon through phosphoglycollate and glycollate. Enzymes of the glycollate pathway are indeed present in bundle sheath cells, but as there is no glycerate kinase in the chloroplasts of these cells (Usuda and Edwards, 1980), any glycerate formed from serine could not be converted to 3-PGA within the bundle sheath.

C_4 species may be classified according to the differences they display in major C_4-acid translocated from mesophyll to bundle sheath cells, the enzymes responsible for decarboxylation, and the C_3-acid returned to the mesophyll cells (Hatch et al., 1975; Huber and Edwards, 1975; Table 6). Irrespective of whether pyruvate or alanine is the C_3-unit returned to the mesophyll cells, ATP is required to regenerate PEP in the reaction catalysed by pyruvate, Pi dikinase. A scheme illustrating reactions and carbon-flow for the NADP-ME sub-group is shown by Kennedy (1986, p.91).

Support for these modes of photosynthetic carbon metabolism has been gained from observations on the properties and distribution of relevant enzymes, especially the two carboxylases. PEP carboxylase behaves as a cytosolic enzyme in the leaves of C_3 species such as spinach (Murray and Bradbeer, 1971) and also in the leaves of C_4 species. Forms of PEP carboxylase typical of C_4 mesophyll cells differ in kinetic parameters and chromatographic behaviour

Table 6.
Sub-Groups of C_4 Species

Initials	C_4-acid transferred	Decarboxylation enzyme	C_3-acid returned
NADP-ME	Malate	NADP-Malic Enzyme	Pyruvate
PEP-CK	Aspartate	PEP Carboxykinase	Alanine
NAD-ME	Aspartate	NAD-Malic Enzyme	Alanine

on anion exchange media compared to forms typical of C_3 mesophyll cells, and also are less sensitive to inhibition by anions (Ting and Osmond, 1973a,b). The high PEP carboxylase activities in mesophyll cells of leaves of C_4 plants, together with the higher affinity for CO_2 shown by PEP carboxylase compared to RuBP carboxylase, account for the very low CO_2 compensation points typical of C_4 species (0-10 ppm, compared to 40-100 ppm for C_3 species at normal O_2 concentrations - see Martin et al., 1972).

In C_4 species RuBP carboxylase is often undetectable in ground mesophyll cells, and confined to the chloroplasts of bundle sheath cells. This selective distribution was confirmed by electrophoretic analysis of extracted proteins (Huber et al., 1976) and immunocytochemical localization of RuBP carboxylase with fluorescent antibodies (Hattersley et al., 1977). Some exceptions to this precise tissue distribution of carboxylases are now being found, e.g. in the genus *Flaveria* (Reed and Chollet, 1985). These odd species are believed to possess metabolic rather than purely spatial determinants of their C_4 mode of carbon metabolism.

Throughout the 1970's a general impression was conveyed that C_4 plants are inherently superior to C_3 plants. This uncritical enthusiasm has now given way to observation and calculation. Brown (1978) proposed that C_4 species could hold two major advantages: -

(1) they are capable of greater dry matter production per unit of water transpired;

(2) they demonstrate greater biomass production per unit of assimilated nitrogen in the plant, i.e. a greater nitrogen use efficiency.

The first of these points is generally accepted. Stomata of C_4 plants can be held at narrow apertures without causing the internal partial pressure of CO_2 to fall below the compensation point, and thus inhibiting net CO_2 fixation, as would occur in C_3 species with their much higher compensation points. With limited water supply, C_4 plants can fix more carbon per unit time.

Ehleringer (1978) has pointed out, nevertheless, that C_4 plants are not necessarily superior in this respect under all combinations of environmental circumstances. He compared the effects of latitude in the Great Plains of North America for one summer month (July) and found that estimates of daily carbon gain were greater for C_3 canopies between 45° and 60° of latitude; C_4 canopies were superior between 25° and 45° of latitude.

A similar comparison made for a 12-month period for the Sonoran Desert (Tucson, Arizona) indicated that C_3 plants would again be superior from about half-way through October until the end of April. Finally, a comparison was made for a shaded habitat (10% of full sun). Estimates of daily net carbon gain plotted as a function of temperature indicated that C_4 plants could be superior to C_3 plants under shaded conditions only at temperatures above 33^oC.

Ehleringer's predictions tallied with observations on the natural flora. For instance, in the Sonoran Desert, all 16 common summer grass species are C_4 plants, whereas two winter grass species are C_3 plants. Ephemerals (herbs and grasses) are almost completely C_3 in the winter (97%), but mainly C_4 in the summer (57%). A general lack of C_4 species in shaded habitats, where temperatures would rarely exceed 32°C, also confirms the disadvantage of a lower quantum yield displayed by C_4 species under such circumstances, where "the increased energy cost . . . can be a significant liability" (Ehleringer, 1978). The conclusion is that the C_4 photosynthetic mechanism may be highly advantageous in specific environments, but disadvantageous in others.

With regard to the claims of nitrogen use efficiency exhibited by C_4 species, Brown (1978) cites several papers purporting to show that the protein contents of leaves of C_4 species are generally lower than the protein contents of leaves of C_3 species. Unfortunately the data in most of these comparative studies are unreliable, having been obtained with the Folin-Ciocalteu reagent (Lowry et al., 1951) and with bovine serum albumin as a reference standard. This assay is subject to interference by leaf phenolics and Tris buffer, and as discussed in detail elsewhere, bovine serum albumin is atypical in this assay, with a lower colour yield than more representative proteins (see Murray, 1986b).

The first true leaf of a young pea seedling shows a maximum extractable protein content which is about 35% of the leaf lamina dry matter content. Its maximal content of RuBP carboxylase (fraction 1 protein) does not exceed 40% of extractable protein (Gordon et al., 1978). In C_3 plants like pea, RuBP carboxylase protein will be present in all the chloroplasts of the mesophyll cells. However, in C_4 species with RuBP carboxylase confined to bundle sheath cells, it is reasonable to expect there to be less protein in mesophyll cells. This indeed was the finding of Huber et al. (1976): C_4 species (*Zea mays* and *Digitaria sanguinalis*) yielded mesophyll chloroplasts with soluble protein to chlorophyll ratios that were about 3-fold lower than the corresponding ratio obtained for mesophyll chloroplasts from a C_3 species (*Nicotiana tabacum*).

It is probably true that C_4 species function with less total protein per unit of leaf dry matter than C_3 species - but earlier estimates should be refined by using more accurate methods of protein determination. The concept is worth pursuing, since nitrogen use efficiency may well be of adaptive value to C_4 plants, but this very efficiency may be contrary to what might be considered desirable for increasing the quantity and quality of protein stored in seeds for human consumption.

3.2 The Synthesis and Export of Sucrose in Leaves

The disaccharide sucrose is the principal form of assimilated carbon translocated from leaves of both C_3 and C_4 species, via the phloem (Chapter 4). Actively photosynthesizing leaves display periodic accumulation and depletion of both sucrose and starch, with maximal contents occurring close together in time, about the end of the photoperiod (Fig. 6). However, it has not been clear whether the starch that disappears in darkness is converted to sucrose, nor whether translocation of sucrose from leaves continues in darkness.

3.2.1 Sucrose Synthesis

Within leaf cells, starch synthesis is confined to chloroplasts, whereas the synthesis of sucrose probably occurs in the cytosol. The precursor is considered to be triose phosphate exported from the chloroplasts, each molecule of sucrose requiring four of triose phosphate (Walker and Sivak, 1985, 1986). The enzymes that catalyse the final synthetic reactions are sucrose phosphate synthase, and sucrose phosphatase, with the activity of the latter often about 10-times the activity of the former (Hawker, 1967, 1985). These enzymes utilize supplies of UDP-glucose and fructose-6-phosphate derived from triose phosphates by the action of aldolase and ancillary enzymes present in the cytosol (Fig. 7).

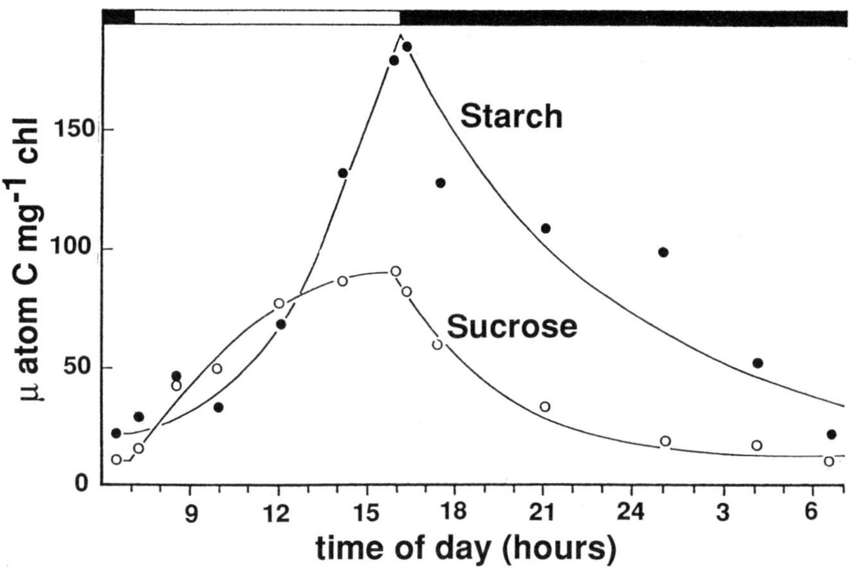

FIG. 6. Diurnal changes in the contents of starch and sucrose in spinach leaves. From Gerhardt, Stitt and Heldt (1987).

* * * * * *

Regulation of the action of enzymes engaged in sucrose synthesis can depend on the concentration of fructose-2,6-bisphosphate, a recently discovered regulatory metabolite (Stitt et al., 1982; Stitt and Heldt, 1985; Huber, 1986). This compound, and the enzymes responsible for its synthesis and breakdown, are confined to the cytosol (Fig. 7). The rate of sucrose synthesis can be slowed by fructose-2, 6-bisphosphate inhibiting the phosphatase that acts on fructose-1,6-bisphosphate (step 3 in Fig. 7). This slows the rate of release of Pi, and will perhaps limit triose phosphate exchange for Pi through the chloroplast envelope (Fig. 7). Conversely, a high concentration of triose phosphate in the cytosol will inhibit the kinase responsible for the synthesis of fructose-2,6-bisphosphate, allowing its concentration to decline, and the rate of synthesis of fructose-6-phosphate to increase.

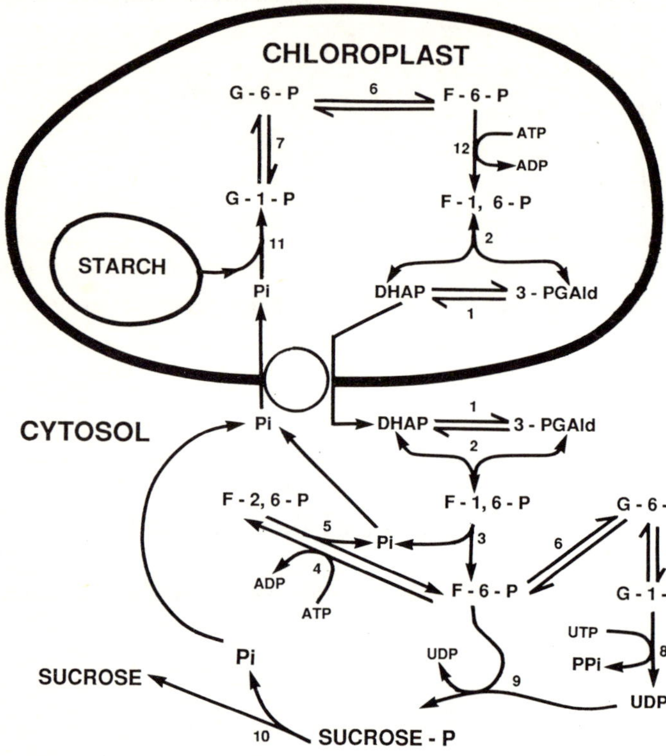

FIG. 7. Hypothetical metabolic scheme linking mobilization of starch within the chloroplast to sucrose synthesis in the cytosol. Enzyme key: 1, triose phosphate isomerase; 2, aldolase; 3, fructose bisphosphatase; 4, phosphofructo-2-kinase; 5, fructose-2,6-bisphosphatase; 6, phosphoglucoisomerase; 7, phosphoglucomutase; 8, UDPG pyrophosphorylase; 9, sucrose phosphate synthase; 10, sucrose phosphatase; 11, starch phosphorylase; 12, phosphofructo-1-kinase.

* * * * * *

An increase in glucose-6-phosphate concentration resulting from the increased formation of fructose-6-phosphate will activate sucrose phosphate synthase (step 9, Fig. 7). This enzyme can be inhibited by Pi, released in the subsequent step catalysed by sucrose phosphatase (step 10 in Fig. 7). This phosphatase can in turn be inhibited by sucrose at high concentrations (about 100 mM), which might come about if the rate of sucrose translocation from the leaf were to decrease. A consequent 'pile-up' of hexose-phosphates would allow more of the regulator fructose-2,6-bisphosphate to be formed, which would inhibit further synthesis of fructose-6-phosphate from fructose-1,6-bisphosphate.

These regulatory aspects of sucrose synthesis are becoming better understood. By comparison, the way in which starch breakdown in the chloroplast is linked to sucrose synthesis is not well understood. Preiss (1982, 1984) has reviewed the regulation of starch synthesis in the chloroplast. A starch synthase that utilizes ADP-glucose as the donor of glucose units to starch is promoted by 3-PGA, but inhibited by Pi, a substrate for starch phosphorylase (11 in Fig. 7). Since this enzyme is unlikely to be involved in starch synthesis (Matheson, 1984) and specific forms are localized in chloroplasts, starch phosphorylase is perhaps responsible for removing glucose units from starch as glucose-1-phosphate.

Conversion of glucose-1-phosphate to triose phosphate for export to the cytosol would require the input of ATP (Fig. 7). Fructose-1,6-bisphosphate produced within the chloroplast could then be converted to triose phosphates by the action of a chloroplast form of aldolase. Both the chloroplast and cytosolic forms of aldolase could participate in both light-dependent and dark phases of carbon metabolism (Anderson and Pacold, 1972).

A ready supply of ATP is available from photosynthetic phosphorylation in the light, but how could an ATP-dependent conversion of starch to triose phosphate be maintained in darkness? Preiss (1984) suggested the need to investigate the possibility that metabolites other than triose phosphate might be transported from the chloroplast in the dark. If glucose-1-phosphate itself, or any hexose or hexose phosphate derivative could be exported from the chloroplast in darkness, there would be no need to invoke any ATP requirement from within the chloroplast. Recently, Stitt (1987) and Servaites et al. (1988) have proposed that a glucose transport system does operate in darkness, obviating the need for an ATP-generating system inside the chloroplast.

If, moreover, there is no new synthesis of sucrose in darkness, the question is raised as to what becomes of the starch mobilized overnight (e.g. Fig. 6). Most released hexose units could be incorporated irreversibly into cell-wall polysaccharides, but this aspect of starch mobilization in leaf cells needs to be confirmed.

3.2.2 Diurnal Rhythms in Translocation

There is a substantial body of evidence that translocation of solutes from leaves is light dependent, and that it will run down and eventually cease in continuous darkness. The idea of a circadian rhythm in translocation of sucrose and other solutes from leaves gained considerable support from studies on sugar cane, *Saccharum* spp. (Hartt, 1965; Hartt and Kortschak, 1964, 1967). For convenience in manipulation, detached leaves were often used, with $^{14}CO_2$ fed to only part of the whole leaf. Translocation of labelled sucrose downwards from the fed region was most rapid in the light, but slowed down if the fed region was darkened.

A number of attempts have also been made to monitor sucrose synthesis and export in the smaller leaves of dicotyledons. Whole leaves are labelled with $^{14}CO_2$ in the light, then the loss of label from the leaf is measured, together with its appearance in other parts of the plant. Advantage has been taken of the well documented occurrence of 'supply' leaves in legumes; each leaf subtending a fruiting node is able to direct most of its translocated assimilate to the fruit or fruits attached at that node (Linck and Sudia, 1962; Blomquist and Kust, 1971; Harvey, 1973).

By providing $^{14}CO_2$ to the leaf subtending the pod of uniform 30 to 35-day old broad bean plants, Kipps and Boulter (1973) recognized two phases of fruit labelling (Fig. 8). The first phase of acquisition of label by the fruit occurred within the photoperiod, and represents the immediate export of ^{14}C-labelled solutes via phloem from the leaf to the fruit. Most of the label entering the pod was allocated to the seeds (Fig. 8). Subsequent analysis showed that sucrose and serine were the major labelled solutes translocated in the light (Kipps and Boulter, 1974).

The second phase in translocation of label to the fruit occurred between midnight and the next morning. The label in the seeds increased about 3-fold in this period, but there was no further decline in the radioactivity of the fed leaf (Fig. 8). Kipps and Boulter (1973) suggested that because the increase in label received by the seeds had not taken place at the expense of labelled solutes in the leaf, it represented the remobilization of labelled sucrose following temporary storage in the stem. In contrast to the distribution of labelled solutes observed in the light, the increased labelling of the seeds in darkness was accounted for by sucrose plus organic acids, with negligible labelling of amino compounds. Sucrose was by far the most abundant labelled solute both in light and dark.

Kipps and Boulter (1973) conducted their $^{14}CO_2$ feeding from about the middle of the photoperiod (12.15 p.m. to 3.15 p.m.). By comparison, Pearson (1974) fed individual leaves of non-fruiting broad bean plants at four different times of the day (which was a 16 h photoperiod plus 8 h dark). Pearson confirmed that the rate of translocation of recently assimilated carbon is most rapid in the light. His data also indicate that only a small proportion of carbon

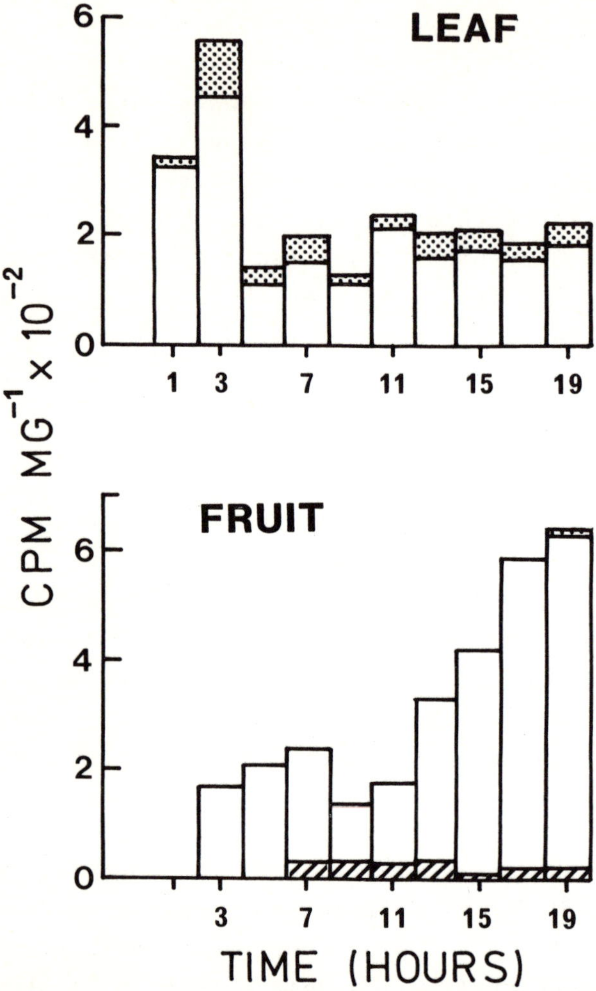

FIG. 8. The distribution of label in the leaf subtending a broad bean pod, and in the pod following photoassimilation of $^{14}CO_2$ by the leaf for 3 h. Within the leaf, ethanol soluble metabolites (unshaded) account for more label than insoluble compounds (stipled). Within the fruit, most of the label was in the seeds (representation as for leaf); little label was in the hull (hatched). Data of Kipps and Boulter (1973).

fixed in the preceding photoperiod can be translocated from a leaf in darkness, and none at all can be removed in the dark period of the second day after feeding.

Sharkey and Pate (1976) used 15-minute labelling periods with $^{14}CO_2$ at five different times within a 10-hour photoperiod. They fed the second leaf below the primary inflorescence of fruiting white lupin plants, then analysed the phloem sap from the tips of the pods (Section 4.2) in the lowest position, to which this leaf directs its assimilates. The specific radioactivity of the sucrose from the pod phloem sap was determined for the first time 2 hours after beginning pulse-feeding with $^{14}CO_2$, then at 4-hourly intervals for about 2 days.

The specific radioactivity of the sucrose was maximal at the initial determination irrespective of the time within the photoperiod that feeding commenced, except for the last treatment, fed 2 hours before darkness. In this case, the specific radioactivity of the recovered sucrose increased to a maximum, then declined after about 6 hours of darkness. With the sole exception of this last treatment, there was a further transient peak in the specific radioactivity of sucrose, set against an overall decline from the initial maximum over the 2-day period of the experiment. This transient increase occurred soonest for the $^{14}CO_2$ feeding treatments with the fewest hours of light remaining before the onset of darkness, and took about 4 hours longer to eventuate in the two treatments fed earliest in the photoperiod.

The timing of the transient increase and decline in the specific radioactivity of the sucrose recovered from fruit phloem sap suggested to Sharkey and Pate (1976) that these peaks could be attributed to the release of ^{14}C-labelled hexose from starch in the chloroplasts of the fed leaves, and its conversion to sucrose available for translocation. They argued that starch labelled earliest in the photoperiod would be surrounded by layers of unlabelled starch, whose prior degradation would be necessary for release of labelled glucose to the precursor pools supplying sucrose synthesis and translocation. The less unlabelled starch deposited, the sooner the labelled starch should be degraded, and contribute to the translocated sucrose. However this hypothesis remains unconvincing if translocation from the leaf slows down and effectively ceases in continuous darkness. A more probable explanation is that the stem provided the labelled sucrose entering the fruit overnight - this conclusion is supported by the observations of Kipps and Boulter (1973) and by other work on stem function (Ho and Peel, 1969; Section 4.4).

The stem receives sucrose from the leaf by day, just as the fruit does. The stem even continues to draw sucrose translocated from leaves when all other possible destinations have been excised (Lovell et al., 1972). The stems of sugar cane (Hawker, 1965) and bean plants (Minchin and Thorpe, 1984) allow a considerable pool of sucrose received from phloem to persist in the cell walls (the apoplast, or free space). In bean stems, for example, this apoplast concentration of sucrose can be 25-60 mM. The most recently received sucrose in the stem is that still in the free space, and this sucrose would be the first available for reloading into the phloem. Reversal from storage to export provides the plant with a means of compensating for the dwindling sucrose output from leaves as a period of darkness continues.

Stem cells with stored sucrose labelled earlier from a pulse of $^{14}CO_2$ in the light would release labelled sucrose to the apoplast and thence to the phloem only after unlabelled sucrose had first been reloaded. This sequence of events adequately explains the labelling data of Sharkey and Pate (1976). Clearly a major function of the stem is to buffer the sucrose concentration in the phloem traces supplying a fruit until the onset of the next photoperiod, when the leaf resumes its role as the primary source of translocated sucrose.

3.3 Assimilation of Nitrogen

The complexities of nitrogen nutrition should not be underestimated. The assimilation of nitrogen from inorganic sources is interwoven with the assimilation of CO_2, directly or indirectly sharing the resources made available by the light-harvesting reactions of photosynthesis. The principal forms of nitrogen imported by fruits are amides or amino acids, compounds representing prior investment of ATP and reducing potential in their synthesis, resynthesis, storage and mobilization. Turnover of nitrogenous solutes is an essential and elementary feature of nitrogen assimilation, and has often been overlooked. It is worth reflecting that any given nitrogenous molecule might be immune from turnover only while in transit between cells, or locked up in the structure of a polypeptide, which itself has only a limited turnover time.

3.3.1 Nitrate

The assimilation of nitrate depends on the concerted action of nitrate reductase (Section 2.2.13) and a nitrite reductase system. Nitrate reductase is frequently composed of two polypeptide subunits, each of approximate MW one hundred thousand (Campbell and Wray, 1983). The enzyme usually prefers NADH as a source of reducing potential, but some forms are known to prefer NADPH (Wallace, 1973; Campbell, 1978). The induction of nitrate reductase activity in plant tissues by nitrate is common. Induction by citrate and other compounds at pH 3.0 has also been observed (Knypl and Ferguson, 1975). Nitrate, rather than NO_2^-, also induces the synthesis of nitrite reductase activity, as demonstrated with a nitrate reductase-deficient mutant of *Arabidopsis thaliana*, which therefore lacks an endogenous source of nitrite (Braaksma and Feenstra, 1982).

Nitrate reductase activity in leaves is modulated by alternating periods of light and darkness, increasing in the light, and declining overnight (Wallace and Pate, 19-65; Sherrard et al., 1979). Nitrate reductase activity is often said to be the rate-limiting step for nitrate assimilation, but such assertions are unconvincing. Canvin and Atkins (1974) showed that the assimilation of $^{15}NO_3^-$ into amino groups was light dependent in leaves, and "ceased abruptly" in darkness. This strong dependence on illumination can be attributed to the requirement for reduced ferredoxin by nitrite reductase inside chloroplasts, and not to any immediate deficiency in nitrate reductase enzyme. Kennedy (1986) has indicated that the nitrite reductase sequence involves the transfer of six electrons from reduced ferredoxin for the production of NH_4^+ and two OH^- per nitrite reduced.

Butz and Jackson (1977) suggested that nitrate reductase inleaf mesophyll cells may co-ordinate the conversion of NO_2^- to NH_4^+ by virtue of its location in the chloroplast envelope. Here nitrate reductase could still utilize NADH from the cytosol, and could release nitrite to the inside of the chloroplast, where it would be reduced stoichiometrically to NH_4^+ in the light. Thus the rate of nitrate reduction could depend on the activity of envelope-localized nitrate reductase. Whether all the nitrate reductase of leaf mesophyll cells is associated with chloroplast envelopes is uncertain, but the hypothesis of combined nitrate reduction with vectorial nitrite transport is attractive, at least for chloroplasts.

It seems unlikely that Butz and Jackson's hypothesis might hold true throughout the plant, since there are circumstances when nitrate transport is not linked obligatorily with reduction. Accumulation of nitrate in vacuoles of leaf mesophyll cells (Waters et al., 1982) or transmission of nitrate to the xylem are two such instances. How nitrate and nitrite reductases might co-operate in non-photosynthetic cells is also not clear.

The measurement of nitrate reductase activity has itself presented many difficulties. Wallace (1986) has provided a valuable review of the relationships between nitrate supply, measured nitrate reductase activity in plant parts, and the extent to which these measurements are reliable indicators of actual nitrate assimilation. Caution is urged "in extrapolating from *in vitro* or *in vivo* nitrate reductase measurements to the actual nitrate assimilation in the root and shoot".

Nevertheless, species can be categorized according to their distribution of nitrate reductase activity in the roots versus the shoots (Pate, 1973; Radin, 1978). At one extreme, plants like the cockleburr (*Xanthium pennsylvanicum*) exhibit no detectable nitrate reductase activity in their roots, and at least 95% of the nitrogen from xylem sap is still in the form of nitrate (Wallace and Pate, 1967). Cereals like oats and maize also exhibit high proportions of total xylem sap nitrogen as nitrate (Pate, 1973) and appear to assimilate most nitrate in the leaves (Wallace, 1973, 1975). In maize, most of the limited nitrate reductase activity of the roots is localized in epidermal cells (Rufty et al., 1986). Once incoming nitrate has moved beyond these cells, it is unlikely to be reduced until after passage through the xylem into the leaves.

Legumes, which normally rely on N_2-fixing nodules on the roots (Section 3.3.3), can augment these endogenous supplies of nitrogen with nitrate, taken up according to the external concentration. If this is too high, then nodules will not form, and the plants will be forced to utilize nitrate as their sole source of nitrogen. Mutants that can nodulate effectively in the presence of high concentrations of nitrate are now being characterized (Carroll et al.

1985; Delves et al., 1987).

So far as the principal location for nitrate assimilation is concerned, no clear pattern among legumes has yet emerged. Often most of the nitrate taken up is translocated via xylem to the stems and leaves. This occurs in members of the tribe Phaseoleae, including soybean (Herridge, 1982a,b), cowpea (Pate et al., 1984), *Phaseolus vulgaris*, *Phaseolus coccineus* and *Vigna radiata* (Wallace, 1986). However, this pattern is not confined to a single tribe, and is found also in peanut (Peoples et al., 1986) and two cultivars of field pea (Wallace and Pate, 1965; Table 7).

Table 7.

The Effect of External Nitrate Concentration on Nitrate Reductase Activities of Root and Shoot Extracts Prepared from Several Legumes During Flowering. Activities are as nmol NO_2^- formed min^{-1} per plant (% of total shown also).

Species & Part	Nitrate supply (mM) 1	2	4	8
Pea[a]				
shoot	27.6 (56)	97.7 (74)	350 (83)	695 (88)
root	21.7 (44)	34.3 (26)	69.6 (17)	93.6 (12)
Chick pea[b]				
shoot	57.0 (49)	84.2 (52)	128.6 (61)	43.3 (47)
root	60.3 (51)	77.8 (48)	83.0 (39)	48.1 (53)
Lentil[c]				
shoot	14.5 (32)	25.8 (42)	68.4 (48)	87.5 (51)
root	31.3 (68)	36.3 (58)	73.1 (52)	84.5 (49)
Broad bean[d]				
shoot	6.6 (4)	97.4 (34)	328.2 (59)	431.5 (52)
root	163.1 (96)	187.2 (66)	229.4 (41)	398.5 (48)

[a]cv. Dundale [b]cv. Tyson [c]cv. ILL30 [d]line 383

Data of Peoples, Sudin and Herridge (1987).

* * * * * *

Lentil and broad bean (both Vicieae) as well as chick pea (Cicereae) all appear to assimilate nitrate about equally well in roots and tops, except at low nitrate concentrations, when lentil and broad bean can assimilate most nitrate in the roots (Table 7). This pattern of nitrate assimilation occurs also in wheat (Wallace, 1986).

Lupins develop most of their nitrate reductase activity in the roots, but at extraordinarily high external nitrate concentrations (>10 mM), nitrate reductase activity develops throughout the plant, except in the embryos of developing seeds (Atkins et al., 1979). In contrast with field pea, the garden pea cv. Greenfeast exhibits most nitrate reductase activity in the roots, at nitrate concentrations in the range between 0.5 and 10 mM (Wallace, 1986). Thus different patterns of nitrate assimilation can occur in different genotypes within a single species.

3.3.2 Ammonium

Ammonium is the product of symbiotic N_2 fixation released to plant cells in root nodules (Section 3.3.3) and the product of nitrite reduction in the plastids of root cells (Dalling et al., 1972a,b) or in chloroplasts (Section 3.3.1). Yet despite the apparent saving in expenditure of reducing potential that NH_4^+ represents, very few plant species prefer to take up exogenous NH_4^+ through their root system (McKee, 1962). Many species can grow only to a limited extent when exogenous NH_4^+ is the only nitrogen source available. This generally results in lower concentrations of other mineral cations translocated upwards, and lower cation contents of roots, as in sunflower (Kurvits and Kirkby, 1980) and *Phaseolus vulgaris* (Chaillou et al., 1986). Higher uptakes of external anions (chloride, sulphate and phosphate) occur in these two species, but this is not evident in tomato plants (Macklon and Sim, 1980).

Ammonium-fed plants of the three species just mentioned cannot obtain sufficient water for normal growth. This may result from a reduced efficiency of ion uptake by the roots, but direct effects of NH_4^+ on membrane permeabilities are suggested as a possibility by Macklon and Sim (1980). A competitive effect that NH_4^+ might exert in xylem vessels is to occupy negatively charged binding sites that might otherwise be available for cation binding and exchange. This could retard transfer of mineral or organic cations from xylem to phloem (Section 4.3).

The other difficulties plants encounter in utilizing exogenous NH_4^+ are most readily explained in terms of charge balance. Excess cation uptake over anion uptake results in acidification of the root environment (see Chapter 7 of Kennedy, 1986). If the pH of culture solutions is continually readjusted to pH 5.8, then soybean plants can grow as well with NH_4^+ as they do with NO_3^- as sole source of nitrogen (Rufty et al., 1983). Experimental modification of external pH allows the plant to conserve bases that might otherwise have been secreted, and to divert carboxylates instead to internal charge balance and effective growth. Presumably legumes and other plants with endogenous sources of NH_4^+ from N_2-fixing symbionts exchange carboxylates and bicarbonate ions with their symbionts, and net losses of solutes to the root exterior are much reduced.

Apart from the delivery of some NH_4^+ to leaves via the transpiration stream, there are several other endogenous sources: the action of glycine decarboxylase in mitochondria (Fig. 4), the reduction of nitrite in chloroplasts, and the action of enzymes such as phenylalanine:ammonia lyase and asparaginase. Usually the action of these enzymes should represent a minor contribution to total endogenous NH_4^+ (Attridge and Smith, 1974; Urquhart and Joy, 1981). Quantitatively, nitrite reduction or glycine decarboxylation will predominate as the principal source of NH_4^+, according to species and circumstances. There are then three clear options for the assimilation or reassimilation of NH_4^+ in leaf cells:

(1) via glutamate dehydrogenase (GDH) in mitochondria;
(2) via glutamine synthetase (GS) in the cytosol;
(3) via GS and GOGAT in the chloroplasts.

Hartmann and Ehmke (1980) demonstrated with mitochondria isolated from pea leaves that the action of GDH could account for the reassimilation of only a minor proportion of the NH_4^+ liberated from glycine. In duplicate treatments from the same preparations of mitochondria, [^{14}C]glutamate and [^{14}C]serine were the only labelled amino acids produced from [^{14}C]citrate and [^{14}C]glycine respectively. But the proportion of glycine-derived NH_4^+ that was reassimilated as the amino group of glutamate was estimated to be at most only 14%. These authors concluded that most of the NH_4^+ produced in leaf mitochondria *in vivo* is lost rapidly to the surrounding cytosol.

Here the cytosolic form of GS might be expected to act, either in light or darkness. Canvin and Atkins (1974) measured significant assimilation of label from $^{15}NH_4^+$ into amino acids of bean leaf discs in darkness, when chloroplast GS could not be operating. From their studies of tobacco mesophyll protoplasts, Neeman et al. (1986) suggested that cytosolic GS and mitochondrial GDH might operate in the dark to generate [^{15}N-amide]glutamine or [^{15}N]glutamate respectively from [^{15}N]glycine. Otherwise, no label from glycine could have become associated with glutamate; cytosolic GS acting alone would transfer label to glutamine, but not to glutamate.

Nevertheless, it is now clear that the combination of GS and GOGAT in the chloroplast mediates the reassimilation of most of the NH_4^+ released to the cytosol by mitochondria as well as any NH_4^+ generated *in situ* from the reduction of nitrite. The chloroplast-localized GS is implicated in the light-dependent formation of glutamine (Lea and Miflin 1974; Mitchell and Stocking, 1975; Givan, 1975; Bauer et al., 1977a,b). It is not surprising to find that NH_4^+ generated locally can be assimilated as glutamine and then glutamate by the GS + GOGAT sequence; in the absence of a high affinity system for removing NH_4^+, photosynthetic phosphorylation might be uncoupled (Wara-Aswapati and Bradbeer, 1974). However, it is surprising to find that the chloroplasts should be responsible for coping with most of the NH_4^+ derived from glycine cleavage in the mitochondria.

The most compelling evidence for this conclusion comes from the characterization and study of mutants of barley and *Arabidopsis thaliana* with specific defects in enzyme activity or metabolite transport (Somerville, 1986; Wallsgrove et al., 1987). Mutants with ineffective GS or GOGAT behave like others with specific lesions in glycollate metabolism (Section 3.1.4). Their inability to grow normally in air highlights the importance of these enzyme activities to the normal turnover of metabolites in C_3 leaf mesophyll cells. With barley mutants defective in GS, NH_4^+ accumulates, demonstrating that the alternative assimilation systems cannot cope *in vivo* (Wallsgrove et al., 1987). Mutants with a defective chloroplast dicarboxylate transporter (Somerville and Ogren, 1983) are functionally equivalent to those lacking GOGAT; they are unable to cycle oxoglutarate and glutamate through the chloroplasts at high enough rates to prevent a build-up of NH_4^+ derived from the decarboxylation of glycine.

In C_4 species, nitrate reduction is evidently confined to the ground mesophyll (Harel et al., 1977; Neyra and Hageman, 1978). Conversely, glycollate metabolism is confined to the bundle sheath cells, accompanying ribulose bisphosphate carboxylase - oxygenase activities (Section 3.1.5). Ammonium ions could be produced in bundle sheath cells from glycine decarboxylase activity, but the serine produced from glycine could not be converted to 3-PGA in bundle sheath cells because the chloroplasts of these cells lack glycerate kinase(Usuda and Edwards, 1980). Re-assimilation of NH_4^+ can take place in the mesophyll and in the bundle sheath, since the assimilatory enzymes GS and GOGAT are distributed throughout these tissues (Harel et al., 1977; Neyra and Hageman, 1978).

The capacity of chloroplasts to generate glutamine from NH_4^+ is essential for the cycling of amino groups within leaf cells. Glutamine, glutamate and aspartate may carry amino groups back to the cytoplasm from the chloroplasts via the dicarboxylate transporters mentioned previously. Labelling studies with ^{15}N or ^{13}N have indicated wide species differences in the way that amino groups are re-distributed within leaf cells, and then made available to the phloem. In *Datura stramonium* glutamine remains the most prominent recipient of label from $^{15}NO_3^-$, both within the leaf, and among the spectrum of nitrogenous solutes translocated via phloem through the leaf stalk (Fig. 9).

In species like barley, glutamate and glutamine retain most of the label from $^{13}NH_4^+$ and contribute substantially to phloem-mediated export of nitrogenous solutes (Hanson and Tully, 1979; Tully and Hanson, 1979). In pea leaves, glutamate and alanine are the major recipients of label from $^{15}NO_3^-$ (Bauer et al., 1977b). In all of these species, serine carries disproportionately more label from newly-fixed carbon than from newly-assimilated nitrogen, whereas the converse is true of translocated glutamine (Fig. 9). This reflects continued import of unlabelled glutamine by leaf cells from the roots and stem (Section 4.4.4). Phylogenetic and developmental differences in the identities and relative abundance of metabolites carrying both nitrogen and carbon from leaves to developing fruit walls and seeds can be anticipated, and need to be fully documented.

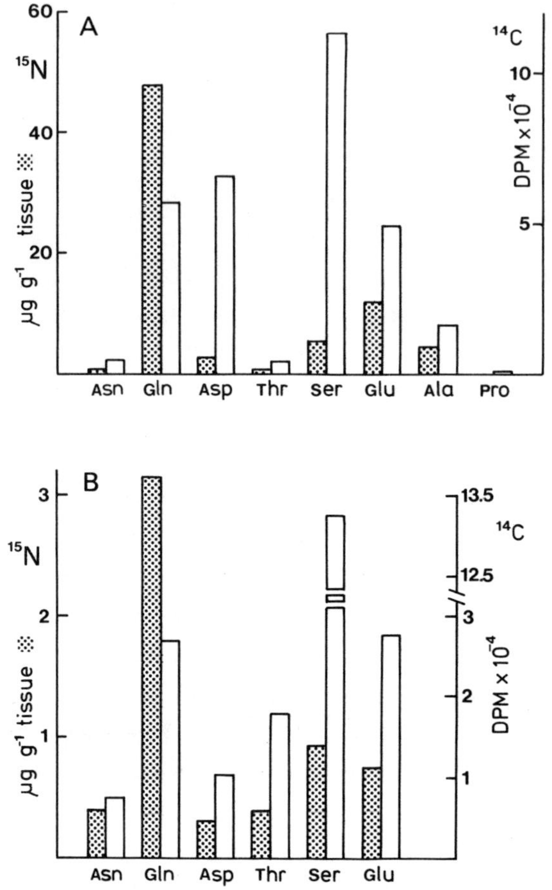

FIG. 9. Incorporation of label from $^{14}CO_2$ and $^{15}NO_3^-$ into the most abundant free amino compounds of leaf blades (A) and petioles (B) of *Datura stramonium*. From Lewis (1975).

* * * * * *

3.3.3 N_2 (Dinitrogen)

The development of nodules enclosing populations of N_2-fixing prokaryotes allows some plants to use NH_4^+ as an 'endogenous' source of assimilated nitrogen (Nutman, 1976; Bauer, 1981; Bergersen, 1982). The precise energy requirement of nitrogenase is not known, but lies in the range 12-20 ATP molecules per molecule of N_2 reduced to two NH_4^+ (Kennedy, 1986).

Transformation of *Rhizobium* infection sites into fully functional nodules on legume roots is accompanied by increases in the activities of host assimilatory enzymes. As in chloroplasts, GS (Section 2.3.2) and GOGAT are responsible for converting NH_4^+ to amide and amino groups, available for transport directly as glutamine, or after conversion to alternative solutes. Asparagine is the most abundant nitrogenous solute exported from nodules via xylem in many species (e.g. Fig. 10), and is derived from a specific pool of aspartate through the action of glutamine-dependent asparagine synthetase (Streeter, 1973; Snapp and Vance, 1986).

Sprent (1980) and Schubert (1986) have divided cultivated annual legumes into two groups according to the vascular anatomy of the nodules, and the major products of NH_4^+ assimilation in the nodules. Temperate 'amide exporters' (examples in Fig. 10) may be contrasted with subtropical or tropical species such as soybean and cowpea, which export the ureides allantoin and allantoic acid as major products of NH_4^+ assimilation. This division into amide exporters versus ureide exporters should not be oversimplified. Ureide producers do not export ureides exclusively, and both asparagine and glutamine are normally major solutes accompanying ureides in xylem sap, in soybean (McClure and Israel 1979) and cowpea (Herridge et al., 1978; Pate, 1984).

The purine-mediated pathway of ureide production is the subject of active research at present (Atkins et al., 1982, 1984; Nguyen, 1986; Schubert, 1986). There are conflicting reports over the occurrence of ureides in xylem sap from nodulated pea, chick pea and peanut plants. Pate (1973, 1977, 1984) has claimed that ureides form a signif-

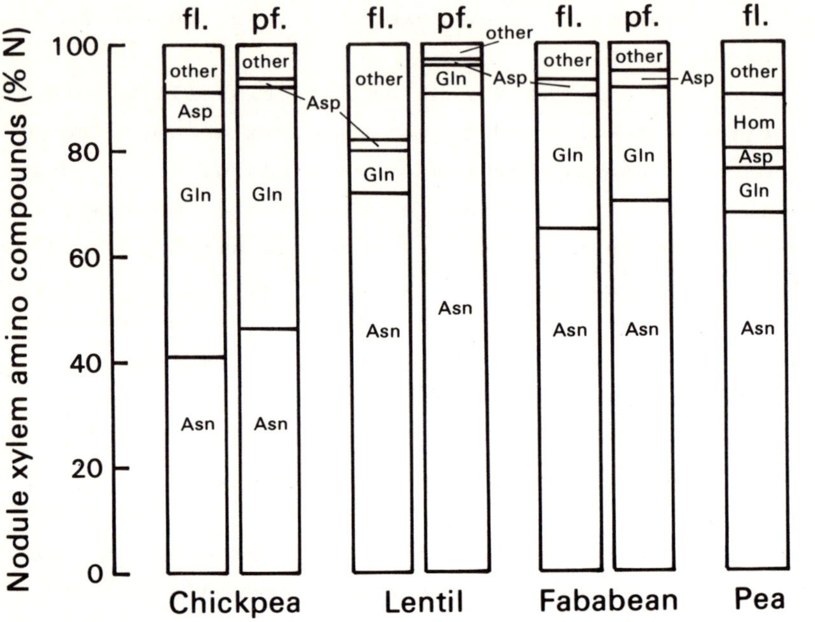

FIG. 10. The distribution of nitrogenous solutes in xylem sap from detached nodules from symbiotic chick pea, lentil broad bean and pea plants. Hom = homoserine; 'other compounds' did not include ureides. The sap samples were collected at flowering (fl) or during pod-filling (pf). Data of Peoples, Sudin and Herridge (1987).

* * * * * *

icant proportion of the total nitrogen in the xylem sap from nodulated pea plants, but are absent from nitrate-fed plants. However, ureides were not detected in the xylem sap from nodulated pea plants by Van Beusichem (1983) or by Peoples et al. (1987), nor from chick pea (Fig. 10). Moreover, it has been shown that the nodules of at least one line of pea (cv. Bodil) should be incapable of ureide synthesis, because they are deficient in two enzymes in the biosynthetic pathway - 5'-nucleotidase and uricase - compared to soybean nodules (Christensen and Jochimsen,

1983). Neither ureides (undetectable) nor 4-methyleneglutamine are significant products of $^{15}N_2$-fixation in peanut - asparagine is the major labelled product (Peoples et al., 1986).

In annual legumes where fruit development is accompanied by the progressive senescence of the whole plant, remobilization of nitrogen from the substantial reserves of the leaves, stalks and stems contributes to the reserves of the fruit wall and the seeds. The extent to which current N_2-fixation contributes to the nitrogen supplies of the seeds varies markedly with species. Symbiotic pea plants derive little nitrogen from fixation during seed development, since "degenerative changes in the nodule population coincide with fruit setting" (Pate, 1977). Chick pea resembles pea in this regard (Sheldrake and Saxena, 1979), and may fix as little as 5% of total assimilated nitrogen during seed development (Hooda et al., 1986).

Studies with $^{15}N_2$ have shown that soybean seeds acquire an increasing proportion of the products of current N_2 fixation as they develop, up to at least 84% (Warembourg et al., 1982), although overall, remobilized nitrogen contributes more than half the nitrogen acquired by the seeds. Cowpea resembles soybean in maintaining functional nodules during at least the early part of seed development (Peoples et al., 1983), then nitrogen fixation falls to zero during mid-fruiting (Herridge and Pate, 1977).

Unlike species that remobilize high proportions of previosly assimilated nitrogen, symbiotic annual peanut plants show an increase in the number and dry matter contents of nodules _after_ the commencement of pod-filling, and maintain increased rates of N_2 assimilation while the fruits are developing (Santos and Sutton, 1982). Substantial removal of foliage up to 10 weeks from sowing has no adverse effect on seed yield. Peanut thus resembles broad bean, the subject of the classic 1880 study by Emmerling [for reproduction, see p.203 of Salisbury and Ross (1978) or p.263 of Salisbury and Ross (1985)]. These data clearly demonstrate a substantial contribution to seed reserves of nitrogen from N_2 fixed after flowering, with a concomitant increase in stem nitrogen rather than a decline.

With further information it will be possible to place annual (monocarpic) species more clearly on a spectrum between garden pea and peanut. The extent to which symbiotic species rely on concurrent N_2 fixation during seed development has implications for the allocation of sucrose from photosynthesis in leaves and stipules of fruiting plants, since those plants which curtail N_2 fixation soonest might, in theory at least, redirect assimilates more economically away from roots to fruits and seeds.

CHAPTER 4
Translocation Mechanisms

4.1 The Physiology of Phloem

The translocation of solutes in plants can be symplastic or apoplastic. Symplastic transfer involves the movement of solutes through plasmodesmata, thin strands of cytoplasm passing through cell walls and connecting the protoplasts of adjoining cells (Gunning and Robards, 1976). Apoplastic transfer involves diffusion of solutes in solution through cell walls or free space (Hawker, 1965).

The specialized vascular tissues enabling long-distance transfer of water and solutes in plants extend these two fundamental modes of intercellular transfer. Phloem sieve elements are connected end-to-end via sieve plate pores, which develop from plasmodesmata (Evert, 1984; Esau and Thorsch, 1985). Symplastic connections are also observed between sieve elements and adjacent parenchyma and companion cells. Mature xylem vessels are dead cells with thickened and lignified walls. They function as fluid-filled capillaries, releasing solutes to various stem cells, or to the apoplast of leaf mesophyll cells in response to vapour pressure deficits in the gas space of the leaf (see Milburn, 1979).

The movement of water and solutes through series of vessels requires no further input of metabolic energy, whereas the movement of solutes via sieve tubes requires living phloem tissues and an input of metabolic energy, as shown by studies involving steam-girdling, or the provision of metabolic inhibitors such as cyanide (e.g. Ho and Mortimer, 1971; Guardiola and Sutcliffe, 1972).

Translocation of solutes within phloem sieve tubes is essentially a physical process; the energy requirements implied by the effects of respiratory inhibitors are mainly concerned with the uptake or loading of sucrose and other solutes (Section 4.4), as well as their 'unloading' some distance away within sink tissues. The contemporary view of how phloem functions is based on the reformulated pressure flow hypothesis according to Münch (1930), which has its beginnings in the work of Hartig last century. The following quotations from Sachs (1875) are illuminating:

> "In the system of communicating sieve-tubes . . . the movement of the substances is necessarily one of the entire mass, caused by inequalities of pressure"

> "That the contents of the perforated sieve-tubes . . . are also subject to considerable pressure from the surrounding tissue is shown by the extent to which these fluids flow out when the organ is cut through."

In the last hundred years alternative hypotheses have been generated in abundance. Biddulph and Cory (1960) suggested that there might be a mechanism in addition to mass flow, but most of the alternative ideas were put forward in response to the erroneous assumption that the sieve-plate pores are normally occluded or blocked *in vivo* (see Swanson, 1959; Biddulph, 1959; Zimmermann, 1960, 1969; Wardlaw, 1974). Zimmermann (1969) quipped that " . . . it sometimes appears as though there are as many concepts as workers", and Leopold and Kriedemann (1975) summed up the situation as presented to many undergraduates thus:

> "The prospect of developing a totally comprehensive theory which accounts for all forms of translocation

and which is acceptable to all protagonists certainly seems remote."

Fortunately this last assessment was unduly pessimistic. Both Mittler (1957) and Zimmermann (1960, 1969) had shown that rates of fluid exudation through an aphid stylet embedded in a single sieve element correspond to a value per second that is three to ten times the volume of a single sieve element. Zimmermann had argued that only mass flow of solution through the connected sieve elements could have maintained such a rate of exudation. The significance of these observations, together with those of Milburn (1970-1972), eluded many reviewers of the mid 1970's.

Objections to the various 'blocked pore' hypotheses were specifically raised by Milburn, who observed that preliminary stem massage desensitized the phloem sealing mechanism in castor bean plants, and led to the exudation of copious amounts of phloem sap following incision. Milburn pointed out that stem massage might be expected to induce blockage of sieve-plate pores if any of the alternative hypotheses were correct, and he defended the simple Münch pressure-flow hypothesis in the following terms (1970):

"(1) In the normal condition plants translocate sugary sap through phloem sieve tubes, the interconnecting sieve plates of which are open. Flow is induced by turgor pressure differences.

(2) Since injury would induce massive loss of assimilated food materials the system is protected by a rapid automatic sealing mechanism which functions so as to block the pores in the sieve plates, which is their main *raison d'être*. It is activated by an unusually high pressure surge of sap through the pores (and probably by other adverse conditions also).

(3) The sensitivity of the sealing mechanism is regulated to suit the plant environment (e.g. wind

> would cause pressure surges in the system) and the translocatory requirements of the growing plant. By balancing the metabolic sealing and anti-sealing components flow is made possible at a rate sufficient for growth."

Milburn (1970) further concluded that the phloem 'pipeline' is elastic, and therefore, on incision, the pressure drop across the sieve plates will be greatest close to the point of incision. The sealing mechanism, involving the well known formation of callose, operates most effectively nearest to an incision, allowing more distant sieve-plate pores to remain open. Rapid exudation from sealed phloem tubes subjected to repeated incision is consistent with open pores and mass flow between sieve elements.

Improved fixation procedures have since confirmed that sieve-plate pores are open *in vivo* (Cronshaw, 1981), and the elastic properties of phloem sieve elements have been measured more accurately (Wright and Fisher, 1983; Kallarackal and Milburn, 1985a). The serious student of translocation should consult the works of Milburn (1970-1979) and his colleagues (Hall and Milburn, 1973; Zimmermann and Milburn, 1975; Smith and Milburn, 1980; Kallarackal and Milburn, 1984, 1985). It now seems incredible that some of the "more tortuous" alternatives to a pressure-flow hypothesis were ever countenanced.

4.2 Analysis of Sap

The various methods employed to obtain samples of vascular liquids have been described by Bollard (1960), Crafts (1961), Zimmermann (1960, 1969), Milburn (1970) and Pate (1976a,b; 1984). Each method has its own particular caveats and problems of interpretation, especially concerning the actual mobility and delivery of detected solutes.

In summary, the sap or exudate attributable to xylem can be obtained by: -

(1) Severing stems of herbaceous plants close to ground level (root pressure method);

(2) Severing petioles (Pate et al., 1964);

(3) Detaching nodules from legume roots (Fig. 10);

(4) Drilling holes into the central trunk of deciduous trees, especially in seasons when the greatest volumes of sap are available;

(5) Centrifugation or volume displacement by pressure of cut stem segments, or more recently, leaf tips (Munns and Passioura, 1984);

(6) Vacuum extraction of cut stems or stalks.

Pate (1976a) once emphasized that "solute concentrations relate strictly to the point of collection", and listed rates of water flux, age or state of plant development, seasonal fluctuation, diurnal fluctuation. plant disease, and selective exchange of solutes (with vessel walls and adjacent cells) as major factors constraining interpretation of data on xylem sap composition.

Phloem sap can be obtained most reliably through the severed stylet of an aphid (Kennedy and Mittler, 1953; Zimmermann, 1969; Fisher, 1987). The volumes of exudate are very small, but this elegant method provides diagnostic criteria against which other procedures for sampling sieve tube contents must be compared. As noted previously (Section 4.1), the precise localization of the stylet in a single sieve element provides reassurance as to the composition of phloem sap, and its replenishment by continuous files of sieve elements.

The problem of cutting the stylet from the body of the aphid has been approached by the application of lasers (Barlow and McCully, 1972; Kawabe et al., 1980) and radio frequency microcautery (Fisher and Frame, 1984), but due to the expense involved, the low yields of sap, and the unattractiveness of many plants to aphids, this high technology approach may not be popular. Fragments of razorblade (Wright and Fisher, 1983) may be more effective than the 80% success rate obtained with one of the laser procedures (Kawabe et al., 1980). An appraisal of difficulties associated with all of these methods, together with a comprehensive list of aphids and their hosts, has been provided by Fisher and Frame (1984).

Incision of bark or other suitable plant parts is the oldest way of collecting sugary exudates originating in the phloem (Crafts, 1961; Milburn, 1970, 1979; Van Die, 1974). Of direct relevance to fruit and seed nutrition, sap samples attributed to phloem have been obtained with difficulty from cut fruit stalks of pea (Table 8), and from the pod tips of representatives of the legume genera *Lupinus*, *Spartium*, *Genista* and *Jacksonia* (Pate et al., 1974, 1975; Sharkey and Pate, 1975, 1976; Hocking et al., 1977, 1978). More recently, a technique termed 'cryopuncturing' has been developed in studies of cowpea fruit development (Pate, 1984; Pate et al., 1984). This involves puncturing the phloem in veins of the fruit stalk (either pedicel or peduncle), or from the dorsal marginal veins of the pod itself, with a needle previously cooled in liquid nitrogen. This treatment apparently injures the resealing mechanism of the sieve-plate pores so that exudation may continue for relatively long times, thus increasing the yield of sap. Cowpea fruits will not produce phloem exudates by cryopuncture if they are younger or older than certain critical stages.

The applicability of cryopuncture to fruits of other species has yet to be fully explored, but it does not work with pea pods (M. B. Peoples, personal communication). However, information from cowpea pod phloem sap analyses and comparison of the two analyses presented for pea in Table 8 suggest a reinterpretation of the earlier data.

Table 8.

Nitrogenous Solutes in Phloem Sap Samples from Fruit-Stalks of Pea Plants (mM).

Compound	Nodulated plants[a]	Nitrate-fed plants[b]
Asparagine	69	62
Aspartate	15	8
Homoserine	22	21
O-Acetylhomoserine	9	-
Glutamine	24	4.5
Glutamate	6	0.1
Alanine	21	4.1
Serine	29	4.8
Glycine	10	0.8
Threonine	4	2.9
Arginine	1	trace
Histidine	1	0.5
Lysine	1	0.9
Proline	-	trace
Valine	-	1.3
Cysteine	-	1.4
Methionine	trace	0.7
Isoleucine	1	0.6
Leucine	trace	0.7

[a] Data of Lewis and Pate (1973) for cv. Meteor

[b] Data of M. B. Peoples for cv. Greenfeast (see Cruickshank et al., 1987). An unidentified ninhydrin-positive compound from this sample might be glutathione

* * * * * *

In particular, *O*-acetylhomoserine, glutamate and glycine now stand out as unlikely to be major nitrogenous solutes of the phloem sap entering the pea pod. *O*-Acetylhomoserine was originally isolated from the hull of pea (Grobbelaar and Steward, 1958) and is also a metabolite of the fruit stalk, acquiring almost as much label as threonine from [U-^{14}C]homoserine withdrawn from a pulse passing to the young pod (Murray, 1985). Both glutamate and glycine are present only in very low concentration in cowpea pod sap obtained by cryopuncture (Pate et al., 1984). Thus all three of these solutes could have been derived mainly from surrounding stalk tissues rather than phloem in the case of the earlier sample (Table 8).

Table 9 lists the partial composition of sap samples obtained from white lupin and cowpea. This comparison includes many of the nutrients traditionally regarded as phloem-immobile (Section 2.4). Manganese has been a focus for concern in the cultivation of lupins, since deficiencies lead to a split-seed disease (Hocking et al., 1977; Snowball and Robson, 1986). *Lupinus albus* (white lupin) is superior to *L. angustifolius* (narrow-leaf or blue lupin) in translocating Mn to its seeds.

Table 9.
Concentrations of Mineral Nutrients Present in Xylem Sap and Phloem Sap from Two Legumes ($\mu g\ mL^{-1}$).

Nutrient element	Xylem[a]	Phloem[b]	Phloem[c]
Potassium	48	1250	950-1360
Magnesium	5	70	400-470
Calcium	14	6	53-107
Iron	0.9	7	80-177
Manganese	2.5	9	34-68
Copper	ND	ND	16-30
Zinc	0.3	5	ND
Boron	1.0	16	ND

[a] sap from cut stumps of field-grown white lupin plants
[b] sap from cut pod tips of field-grown white lupin plants
[c] phloem sap obtained by cryopuncture of dorsal veins of cowpea pods, both nitrate-grown and symbiotic. Data of L. Campbell & J. S. Pate (lupin) and Pate et al. (1984).

Among these 'immobile' nutrients, only calcium appears sometimes to be less concentrated in phloem sap than in xylem sap from the same plants (Table 9; see also Hocking, 1980). This may reflect the adsorption of Ca^{2+} to surfaces exposed by cutting pod tips, since there is no lack of Ca in fruit phloem sap samples obtained by cryopuncture (Table 9), nor in stem exudates from castor bean (Smith and Milburn, 1980a; Hocking, 1982) or *Grevillea leucopteris* Hocking, 1983b). The translocation of calcium in both xylem and phloem is to be expected, probably as a chelate with various organic solutes (Section 4.3).

The sap samples documented in Tables 8 and 9 illustrate many of the general statements made about xylem sap or phloem sap composition. The total solute concentrations in phloem sap are much greater than those in xylem sap, and in most species of plants, sucrose is its most abundant organic solute, often about 20% (w/v). There is little or no sugar in xylem sap except in trees (Van Die and Willemse, 1975). Potassium is the most abundant mineral cation in phloem sap, where its concentration greatly exceeds that found in xylem sap (Table 9). However, there is no clear relationship between sucrose and K^+ concentrations in phloem sap. This ratio can be modified in favour of K^+ by continuous darkness (Sharkey and Pate, 1976; Smith and Milburn, 1980a) or by varying the mineral composition of solutions supplied through the xylem (Hoad and Peel, 1965). In fruiting wheat plants, the content of K^+ in peduncle phloem sap was found to decline substantially in the three weeks following anthesis, a change correlated with increasing ratios of amino compounds (mainly glutamine) to sucrose (Fisher, 1987).

The reported occurrence of free glucose and fructose in significant concentrations in phloem sap has been interpreted as indicating sucrose metabolism in the tissues adjacent to incisions, and the inclusion of derivative metabolites from these tissues in the samples (Pate et al., 1974). For this reason, the inclusion of chelators such as EDTA in solutions used to assist exudation from cut soybean pods (Fellows et al., 1978) or petioles (King and Zeevaart, 1974; Urquhart and Joy, 1981) has been criticised. Nevertheless, Groussol et al. (1986) have argued that provided the optimal chelator concentration and solution pH are first determined for any given species, the phloem mobility of novel compounds like pesticides can be assessed cheaply and accurately. With these precautions, hydrolysis of sucrose can be minimized, and so too can other unsuspected modifications to authentic phloem solutes brought about by the sampling procedure.

Information on the identities and concentrations of the principal nitrogenous solutes found in xylem and phloem saps has steadily been gathered. The concentration range for total nitrogenous solutes is lower for xylem (0.1 to 5 mg mL^{-1}; Pate, 1976a) and the composition is strongly influenced by the plant's principal nitrogen source(s). The nitrogenous solutes appearing in xylem sap are major derivatives of symbiotic N_2 fixation (Fig. 10), or of inorganic nitrogen assimilation in the root system. Proline is also sometimes present (Hocking, 1980). Many of these major solutes are subsequently transferred to phloem (Section 4.4). As well as being transport and storage solutes for assimilated nitrogen, these compounds may well serve to varying degrees as natural chelating agents for mineral cations, a function that might be sustained in phloem as well as in xylem (Section 4.3).

Nitrate ions may predominate over organic nitrogenous solutes in xylem sap, especially if only a trace or minor capacity for nitrate reduction is present in the root system (Section 3.3.1). However, it cannot be assumed that NO_3^- is never transferred to phloem sap. Castor bean plants prefer nitrate over ammonium as nitrogen source (Allen et al., 1985), and when grown with nitrate can yield as much as 3 mM NO_3^- in stem phloem sap (Smith and Milburn, 1980a). White lupin plants grown over a range of external nitrate concentrations exhibited nitrate in phloem sap samples obtained by incision from four different locations, including the developing pod wall (Atkins et al., 1979).

4.3 The Role of Chelates in Mineral Translocation

It is fundamentally important for translocation within the plant for xylem sap to be maintained at an acid pH, whereas phloem sap is often neutral to slightly alkaline. Some examples are given in Table 10.

Table 10.
The pH of Xylem Sap Compared to Phloem Sap.

Species	Xylem sap	Phloem sap	Reference
Ricinus communis	6.0	8.2 (stem)	A
		7.3 - 7.6	B
Nicotiana glauca	5.6-5.9	7.8-8.0 (stem)	C
Lupinus albus	6.3	7.8-7.9 (pod)	D
Spartium junceum		8.0 (pod)	D
Quercus rubra	4.9-5.0	7.5 (bark)	E
Vigna unguiculata		6.5-7.0 (pod)	F

A, Milburn (1972); B, Smith & Milburn (1980a); C, Hocking (1980); D, Pate et al. (1974); E, Van Die & Willemse (1975); F, Pate et al. (1984).

* * * * * *

The lignified walls of xylem vessels are negatively charged; they can attract and retain any positively charged molecular species, such as mineral ions not appropriately sequestered (Shear and Faust, 1970; Wolterbeek et al., 1984). Carboxylate anions belonging to organic acids are not attracted to xylem walls *per se*, and seem to afford the ideal vehicle for divalent mineral cations that might otherwise be completely immobilized. Organic acids such as citrate, malate, malonate and succinate have been recorded as present in xylem sap, and are increasingly supplied to xylem when NO_3^- is absent (Israel and Jackson, 1978; Van Beusichem, 1983). Increased contents of organic acids in xylem sap accompany increased concentrations of both univalent and divalent mineral cations.

Ferguson and Bollard (1976b) showed that ^{45}Ca moved much more rapidly through cut pieces of apple stem when supplied together with citrate rather than aspartate or chloride, and earlier, Millikan and Hanger (1965) demonstrated the phloem mobility of ^{45}Ca supplied as chelates with citrate or EDTA to the leaves of pea and other species. There is also long standing evidence that iron must form chelates in order to be mobile (Bollard, 1960). Iron is normally taken up by the roots of plants in some form of natural chelate with 'siderophores', high affinity Fe^{3+}-chelating agents released by soil bacteria (Castignetti and Smarrelli, 1986). These authors have proposed that nitrate reductase might participate in the reduction of Fe^{3+}-chelates absorbed by root cells. The localization of root nitrate reductase primarily in epidermal cells of at least some species (Rufty et al., 1986) is consistent with this suggestion. Further traffic of Fe^{2+} within the plant would then involve frequent equilibration with a succession of alternative chelating agents.

Organic acids are normal constituents of phloem sap as well as xylem sap, and can attain concentrations of about 20 mM (Smith and Milburn, 1980a). In phloem sap obtained by cryopuncture from cowpea pods, malonate is the most abundant organic acid, and citrate the least (Pate et al., 1984). Mineral chelates with organic solutes have been isolated from the phloem sap of castor bean plants (Van Goor and Wiesma, 1976). Low MW complexes (1,000 - 5,000) of zinc or manganese have been identified. The net charge of the zinc complexes was negative. Both the zinc and the manganese complexes included esterified phosphate.

Recent evidence indicates that asparagine, the most abundant nitrogenous solute of both xylem sap and phloem sap under most circumstances of nitrogen nutrition for many legumes, is well suited to function as a chelating agent for divalent mineral cations. Using Cu^{2+} as a model cation Cruickshank et al. (1987) have confirmed that asparagine forms complexes with divalent metal ions in a molecular ratio of 2:1 (asparagine:metal ion). It is interesting that chelates of divalent metal cations with asparagine (2:1) or aspartate (1:1; Ferguson and Bollard, 1976b) would still possess net positive charge at the pH of xylem sap (Table 10). They would therefore be capable of binding to negatively charged sites on xylem walls, thus facilitating their transfer to adjacent cells.

There is sufficient information to indicate that most mineral cations are normally translocated as complexes with one or more organic solutes, which may be exchanged as opportunity provides. This general principle has taken a long time to emerge (see Bollard, 1960), but now needs to be tested with a wide variety of plant species. With the availability of sampling procedures that will yield sufficient volumes of phloem exudate (Section 4.2), such detailed study is now possible.

4.4 Phloem Loading - Stems versus Leaves

The transfer of solutes and water from xylem vessels to phloem sieve tubes has frequently been observed. Some of the phloem-borne solutes are drawn mainly from xylem; others like sucrose can be drawn from the apoplast following release from stem parenchyma (Minchin and Thorpe, 1984) or leaf mesophyll cells (Giaquinta, 1983; Thorne and Giaquinta, 1984; Secor, 1987). In a normal day as the photoperiod is succeeded by darkness, the stem replaces the leaf as the immediate source of sucrose directed to fruits (Section 3.2.2). Labelled sucrose applied to soybean leaves can be transferred to developing seeds even after the leaves have turned fully yellow and completely lost photosynthetic capacity (Benner and Noodén, 1984). Thus the ability of leaf vein phloem to function persists late into leaf senescence, when the leaf itself limits the availability of solutes for loading.

In experiments where labelled solutes are supplied via the cut stem, it cannot be assumed that all of the label recovered from the fruit wall and the seeds has been transferred to stem phloem directly from stem xylem. Pate et al. (1975) have emphasized the role of the leaf in "attraction, processing and partial re-export of the many classes of solutes which arrive continuously from roots in the transpiration stream". Thus xylem is perceived as a major avenue for the acquisition of nitrogenous solutes by leaves (Pate et al., 1979b; Atkins et al., 1983). This indirect route to fruits via leaves seems to be most important for phosphorus and sulphur, but in my view unimportant for nitrogen, unless NO_3^- is the predominant source, and most of the plant's nitrate assimilation is occurring in the leaves (Section 3.3.1). Even so, the supply of NO_3^- from the roots may limit nitrate assimilation in leaves within 2 weeks of flowering (Nelson-Schreiber and Schweitzer, 1986), and the export of nitrogen from leaves is then supported increasingly by the breakdown of leaf protein (Waters et al., 1980; Peoples et al., 1980, 1983).

4.4.1 Phosphorus

Certain species exhibit slow lateral transfer of phosphate compared to rapid upward translocation to the leaves. This can be highly specific, even to longitudinal halves of the individual leaves, as in peppermint (Rinne and Langston, 1960). Linck and Swanson (1960) demonstrated with pea (cv. Alaska) that a leaf subtending a developing pod could translocate up to 90% of the ^{32}Pi applied to the leaf surface to the pod within a few hours of the application.

That this translocation was mediated by phloem was confirmed by steam-girdling the petiole of the leaf, when the transfer of label to the pod was inhibited by more than 99.5%. Neumann and Noodén (1984) supplied ^{32}Pi to the cut stem of fruiting soybean plants, then followed the changes in the distribution of label in the plants over 48 hours. In the first hour, 75% of the ^{32}P accumulated in the leaf laminae, with 21% in the stem and petioles, and only about 2% in the pod. After 48 hours, the pod had acquired 38% of the label, with 30% in the seeds.

Leaves clearly become important sources of P translocated to developing fruits in legumes. Nevertheless, it must be noted that xylem to phloem transfer of P commonly takes place in the stem, in herbaceous and woody plants alike (Gustafson and Darken, 1937; Bieleski, 1973). The extent to which leaves receive phosphorus in phloem rather than in xylem needs to be determined for more species, but appears to be widespread.

4.4.2 Sulphur

The assimilation of SO_4^{2-} in plants is believed to take place primarily in leaves, where the requirements for ATP and reducing potential can be met by photosynthesis (see Kennedy, 1986). Not all the sulphate acquired by plants is reduced to the level of the thiol group of cysteine: SO_4^{2-} itself is an essential anion (Section 2.2.6). The tripeptides glutathione (γ-glutamyl-cysteinyl-glycine) and homoglutathione (γ-glutamyl-cysteinyl-β-alanine) are mobile forms of assimilated sulphur synthesized in leaves (Macnicol and Bergmann, 1984). Low concentrations of cysteine and methionine are also found in phloem sap (Table 8).

Sulphate is mobile in both xylem and phloem. Transfer of $^{35}SO_4^{2-}$ from stem xylem to stem phloem was shown to occur in bean plants by Biddulph and Cory (1960). This transfer took place mainly at the nodes. Thus leaves assimilating sulphate are likely to receive their supplies mainly from

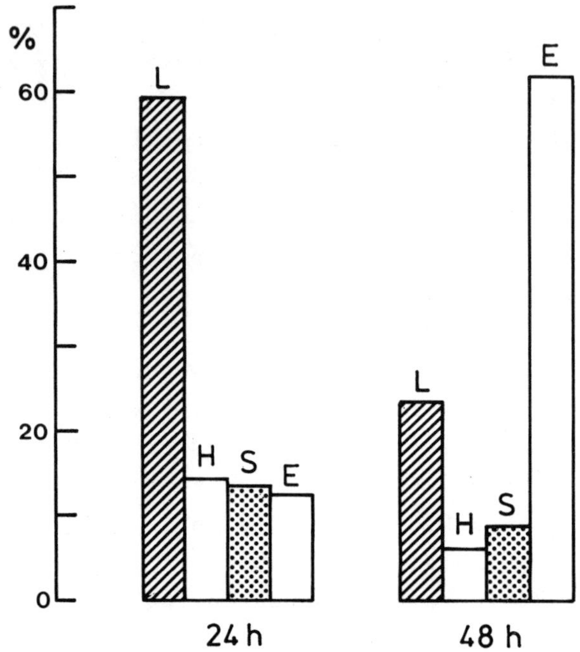

FIG. 11. Distribution of label (as % of total recovered) in the leaves (L), hull (H), seedcoats (S) and embryo (E) of symbiotic chick pea shoots bearing a single pod 24 h and 48 h after feeding $^{35}SO_4^{2-}$ via the cut stem. Data of Sastry and Murray (1987).

* * * * * *

the phloem rather than the xylem, whether the sulphate has been taken up by the plants via the roots or through the cut stem.

When $^{35}SO_4^{2-}$ was supplied to the cut terminal leaf of mung bean plants over a 2.25 h period, the label recovered in organic forms was highest in the peduncle and fruit (Macnicol and Bergmann, 1984). Most of the assimilated label recovered from the leaf laminae was in γ-glutamyl-cysteine whereas most recovered from the peduncle and fruit parts was in homoglutathione. When $^{35}SO_4^{2-}$ was supplied through the cut stem to fruiting fronds of chick pea plants, each with a single pod, labelling of the embryo was more pronounced 48 h from the beginning of feeding (Fig. 11). At this stage, about 80% of the label in the embryo had been incorporated into reserve proteins, predominantly the albumins (Sastry and Murray, 1987). Labelling of the embryo proteins appeared to take place at the expense of the label in the leaves (Fig. 11). Although this trend may have been exaggerated in this study because the stems were not included in the material analysed, the data nevertheless support the view that the leaves are chiefly responsible for the primary assimilation of sulphate, which is then translocated as tripeptide-bound cysteine via phloem to developing fruits and seeds. Sulphate alone is insufficient for the growth of the embryo (Section 8.2).

4.4.3 Other Minerals

McNeil (1980) has argued that the stem of lupins loads K^+ and Mg^{2+} into phloem independently of cation cycling from the leaves, thereby enriching the phloem traces serving the developing fruits. This conclusion is supported by studies of both anion and cation acquisition by developing tomato fruits, supplied via the stem, with or without associated leaves (Wolterbeek and de Bruin, 1986a,b). These studies, together with those of Van Bel (1984), indicate that xylem to phloem transfer of solutes in the stem is inversely related to the rate of transpiration, i.e. the rate of water movement through xylem vessels. Thus solute transfer from xylem to phloem in the stem should be enhanced at night, while the transpiration rate of the plant is minimal. This accords with observations on diurnal changes in mineral concentrations in xylem sap from white lupin plants, where minimum values are recorded at night (Hocking et al., 1978).

Within stems, particularly in dicotyledons, solute transfer to phloem is highly directed, and often confined to the same vascular trace, as in cottonwood (Fisher et al., 1983). Transfer may be concentrated at the nodes, and mediated by parenchyma (cottonwood), or by 'transfer cells' with obvious wall invaginations, which have the effect of enlarging the surface area of the plasmalemma (Pate et al. 1970; Gunning et al., 1970).

4.4.4 Nitrogenous Compounds

Nitrogenous solutes introduced to the xylem through the cut stem may behave differently from one another in the same plant, or their fates may vary in different species. In white lupin, ^{14}C-labelled asparagine and valine are transferred from stem xylem to fruit phloem with practically no conversion to other translocated solutes (Sharkey and Pate, 1975; McNeil et al., 1979). Other amino compounds such as glutamine, threonine and serine are partly (25-50%) converted to other translocated solutes. A final group, including alanine, aspartate and glutamate, is made up of compounds largely transformed by metabolism in intervening tissues to different metabolites, including sucrose and organic acids. Homoserine is not a major translocated solute in lupin, but when supplied through the cut stem in this species, it conforms to the almost completely metabolized group of compounds. In fruiting symbiotic pea plants, homoserine is strongly transferred to phloem supplying the fruit (Section 5.3).

The amino compounds undergoing the greatest degree of alteration en route to developing fruits might have been converted to other metabolites within the mesophyll of 'supply' leaves. These derivatives could then have been loaded into phloem and exported from such leaves. Alternatively, the metabolic transformations might have occurred mainly in stem cells, then the derivative compounds could have been loaded into phloem within the stem, especially overnight (Section 3.2.2).

Table 11.
Effects of Removing the Subtending Leaf on Recovery of Label in the Pea Fruit 24 h After Pulse-Feeding ^{14}C-Labelled Amino Compounds Via the Cut Stem (% of Total Fed).

Labelled solute supplied via stem	Leaf attached	Leaf removed
[U-^{14}C]asparagine	6.92[d]	6.85[b]
[U-^{14}C]alanine	13.20[a]	9.97[b]
[4-^{14}C]aspartate	9.78[b]	3.13[b]
[3-^{14}C]serine	7.99[c]	10.95[b]
[U-^{14}C]glutamine	8.95[d]	16.94[d]

Data compiled from the author's published and unpublished results for cv. Melbourne Market, 1981-1986 spring seasons (for methods see Murray [1983b, 1985, 1986a], Murray & Cordova-Edwards [1984a,b] and Murray et al. [1971]. Ages of pods: a, 14-15 days; b, 15-16 days; c, 16-17 days; d, 18-19 days).

* * * * * *

Pulse-labelling experiments with uniform symbiotic pea plants bearing a single pod of known age have been performed with the subtending leaf attached, or cut off at the moment feeding is commenced via the cut stem (Table 11). Removal of the supply leaf deprives the developing pod of the sucrose and amino compounds normally translocated from the leaf via phloem, as well as the opportunity to acquire labelled solutes or their derivatives through 'leaf cycling'. All label gained by the pod under these circumstances will have come by transfer of solutes to phloem in the stem. Two components can be recognized - direct xylem to phloem transfer, and a 'stem cycling' contribution that might be important overnight.

In the case of asparagine, the nitrogenous solute in

Table 12.

Distribution of Radiactivity Among Soluble Metabolites of the Hull 24 h After Pulse-Feeding [U-^{14}C]asparagine to Fruiting Pea Plants Via the Cut Stem (% of Total).

	Supply leaf attached[a]		Leaf severed[b]
Compound	16-17 d	18-19 d	15-16 d
Alanine	1.00	0.52	0.83
Aspartate	1.26	0.44	0.31
Asparagine	69.22	88.70	77.33
Sucrose	7.68	5.30	5.79
Glutamate	1.17	0.31	2.05
Glutamine	3.27	0.62	3.68
Serine	2.51	0.30	1.19
All others	13.89	3.86	8.82

Pod ages as days (d) after flowering; [a] 1981 season; [b] 1985 season. Data of Murray & Cordova-Edwards (1984b) and Murray (1987).

* * * * * *

highest concentration in phloem sap entering the pod (Table 8), removal of the subtending leaf makes little difference to the amount of label recovered in the fruit (Table 11). Since most of the label in the peduncle (ca. 70%) and the hull (Table 12) is still associated with asparagine itself, it may be concluded that phloem loading of asparagine directed to the fruit takes place almost entirely within the stem.

Excision of the supply leaf diminishes the label recovered in the fruit when alanine or aspartate is fed (Table 11). This positive contribution of the subtending leaf, most substantial in the case of aspartate feeding, is consistent with this leaf being responsible for significant metabolism of the fed solute, then export of derivatives, especially sucrose. Labelled sucrose is the predominant derivative secreted into the embryo sac of young pea seeds

6 h after stem-feeding [4-^{14}C]aspartate to a plant with the subtending leaf attached; most of the label recovered in the hull, seedcoats, embryo sac liquid and embryos 24 h after feeding is in sucrose.

The most prominent labelled products recovered from pod components after [U-^{14}C]alanine feeding with the supply leaf attached are sucrose and glutamine (Murray, 1986a). However, the proportions of label in these two metabolites shift in favour of glutamine when the supply leaf is cut off, indicating that this leaf is responsible for some, but not all, of the conversion of alanine to sucrose undergone en route to the pod. The stem is also involved in the conversion of [U-^{14}C]alanine to sucrose imported by the fruit, because a detached pod fed via the peduncle converts very little alanine to sucrose; instead, glutamine is the major labelled derivative secreted to the embryo sac. Thus the hull has been identified as the major site for conversion of imported alanine to glutamine then transmitted to the seeds, whereas normally both the stem and the subtending leaf contribute to the conversion of alanine to sucrose, and its loading into phloem directed to the pod.

The conversion of [3-^{14}C]serine to sucrose following stem-feeding is also substantial. With the subtending leaf attached, about 25% of the label among solutes secreted into the embryo sac is still associated with serine; the major secreted derivatives are sucrose (40%) and glutamine (14%). In the absence of the subtending leaf, normally the principal source of serine for the fruit (Section 3.1.3), the conversion of labelled serine to sucrose is far more extensive: only 5% of the label in solutes entering embryo sac liquid is still associated with serine, with a similar proportion in glutamine (6%). Most of the label (ca. 70%) is in sucrose. These observations emphasize the general usefulness of serine throughout the plant - with a higher proportion of a pulse of labelled serine surviving to enter the embryo sac only when a plentiful supply of unlabelled serine is available to the same fruit from the subtending leaf.

When labelled glutamine is supplied to fruiting pea plants via the cut stem, most of the label subsequently recovered is present about equally in the leaves and stems (Lewis and Pate, 1973). Twenty four hours after the provision of [^{15}N-amide]glutamine, the hulls and seeds from pods at four or five fruiting nodes together contained only 18.66% of the recovered label (Lewis and Pate, 1973). About 80% of this label had been acquired by 10 h from the commencement of feeding, implying prompt delivery through stem phloem, with little remobilization of labelled derivatives of glutamine from the stem overnight.

In pea plants with a single fruit at the first flowering node, the pod accounted for only 3.45% of the label fed at the youngest stage of development studied (12-13 days, Murray and Cordova-Edwards, 1984a). The proportion of fed label recovered in the fruit increased to about 9% as the pod developed (as in Table 11). However, in the absence of the subtending leaf, the labelling of the fruit approximately doubled (Table 11).

Thus the results of 'minus leaf' experiments with both [3-^{14}C]serine and [U-^{14}C]glutamine confirm that the subtending leaf behaves as a competing sink, and imply once again that the selective import of nitrogenous compounds by the leaf is via phloem rather than xylem.

4.4.5 Conclusions

Leaf-cycling is clearly important in some species for the assimilation of phosphate and sulphate, and their delivery to fruits via phloem. The leaves are also important in conducting the primary assimilation of nitrate (Section 3.3.1). Nevertheless, studies of pea plants from which the supply leaves have been excised establish that the translocation of asparagine to fruits depends almost entirely on its transfer from xylem or stem storage cells to phloem within the stem. Translocation of glutamine to the fruit also depends substantially on xylem to phloem transfer within the stem, although some glutamine can be generated in the subtending leaf, or in the hull and seedcoats (Section 6.3.3).

Even where other amino compounds have been judged to pass appreciably through leaves en route to the fruit, it appears probable that they or their immediate derivatives have first entered the leaf via phloem rather than xylem, given the abundance of xylem transfer cells at departing leaf traces in legumes. McNeil et al. (1979) concluded that the leaves of white lupins preferred to take up nitrogenous solutes which had "escaped the stem uptake system", such as aspartate and glutamate. They considered this uptake to be from xylem. Against this, my own evidence for symbiotic pea plants indicates that leaves prefer to import glutamine and serine as nitrogen sources, rather than asparagine, homoserine, or even alanine *per se* (Murray, 1986a). Serine, glutamate and threonine are the most abundant amino compounds accumulated in leaves subtending pods as their maximal free amino nitrogen content is attained (between 2 and 3 weeks following anthesis of the subtended flower). At least two pools of serine can be identified: one pool labelled strongly from [U-^{14}C]alanine appears to be unavailable for immediate export, and thus distinct from the translocated pool generated by photoassimilation of CO_2 (Sections 3.1.3, 3.3.2).

There is no evidence to support the view that amino compounds delivered to the apoplast of leaf mesophyll from xylem vessels in terminating leaf veins can contribute sub stantially to the nitrogen economy of the leaf, or to the import of nitrogenous compounds by fruits.

CHAPTER 5
Fruit Development

5.1 Patterns of Development

The growth of any given fruit does not occur at a constant rate, nor even at a constantly increasing rate, whichever parameter of growth is being measured. Increases in fresh weight, volume, and all the components of dry matter content show characteristic periods of maximal increase and decline. Often the growth of the fruit as a whole seems to slow down as seeds begin to grow most rapidly. Information on fruit growth, development and composition has mostly been sought for fleshy fruits - drupes, berries or other compound types - of obvious economic importance. Changes in sugar content (Section 5.4), acid content, vitamin content, pigmentation, aroma, strength of the cell walls, and the enzymology of these processes are well described for fleshy fruits (Bollard, 1970; Ferguson, 1984; Hawker, 1985, Brady, 1987). There is less information on fruits that are natural food sources for animals and birds (Howe, 1986), and on fruits that dry out at maturity - follicles, pods (legumes), cereal grains and capsules of various types.

It is now increasingly recognized that the sequence of changes in developing fruits is genetically determined, and the whole process of fruit ripening can provide a useful experimental system for the study of gene expression (Brady, 1987). With this background in mind, it is the purpose of this chapter to examine the relationships between fruit vascular arrangements and nutrient supply (Section 5.2), to assess the extent to which green fruits are autotrophic for carbon (Section 5.3), and to enquire whether nutrients previously committed to fruit walls can be redirected to seeds as dry fruit types mature and desiccate (Section 5.5).

5.2 Vascular Relationships and Phloem Unloading

5.2.1 Nutrient Supply

There is at present no consensus concerning the extents to which the xylem and the phloem directly contribute to the nutritional requirements of fruits. Münch (1930) and Mason and Maskell (1931) proposed that phloem is the predominant if not the sole avenue of nutrient supply to developing fruits. Crafts (1961) and Bollard (1970) have also provided evidence supporting the view that weakly transpiring organs such as fruits obtain most of each single nutrient from phloem sap. Bollard (1970) pointed out that increases in the mineral contents of stone fruits occur in concert with dry matter accumulation, i.e. mineral import often accompanies the acquisition of sucrose, which is imported entirely in the phloem.

Canny (1973) demonstrated extensive development of secondary phloem in the fruit stalk, accompanying fruit development in the sausage tree, *Kigelia pinnata*. For the purpose of calculating specific mass transfer, Canny assumed that the import of nutrients contributing to dry matter increase in the fruit was entirely via phloem.

Specific mass transfer is a measurement of the amount of material translocated per unit time per square cm of sieve tube cross-sectional area. Using improved methods of distinguishing sieve elements from other phloem cell types, Kallarackal and Milburn (1984) compared specific mass transfer values attained by phloem supplying fruit clusters versus single fruits of castor bean. The phloem serving a single fruit achieves a specific mass transfer value only about double that typical of fruits developing in competition with one another - 12.12 compared to 6.18 g h^{-1} cm^{-2}. Analysis of phloem sap from peduncles confirmed that sucrose was the predominant constituent (87.5% of dry matter at the stage studied).

The assumption that practically all a fruit's nutritional needs are delivered from phloem gains further support from observations on the peanut, *Arachis hypogaea*. After flowering, fruiting stalks elongate into gynophores, which arch into the soil around the plant. Peanut fruits develop underground; they can draw no assimilates by transpiration and conduct no photosynthesis. There may be localized uptake of Ca^{2+} and SO_4^{2-} from the pegging zone by hair-like roots on the gynophores (Harris, 1949). Although the mode of delivery of these two nutrients following uptake via the gynophore has not been established, it is likely to involve the phloem eventually. With these possible exceptions, the entire nutrient supply of the peanut fruit must come through phloem from the parent plant.

The composition of phloem sap obtained from pod veins by the most reliable methods (Table 9) clearly demonstrates the ability of plants to funnel all essential nutrients to fruits through phloem. The solitary exception to this may be silicon (Section 2.2.15). Even supposedly immobile nutrients such as calcium, iron, manganese and boron are concentrated in fruit vein sap samples obtained by the methods least susceptible to sampling artifacts (Pate et al., 1984). A variety of mineral ions, nitrogenous solutes and other metabolites, probably as chelates (Section 4.3), must therefore be transferred from xylem or tissue reserves to phloem wherever such opportunities for transfer occur in the vegetative parts of the plant (Biddulph et al., 1961; Section 4.4).

Despite the accumulated evidence of more than 60 years that phloem is the predominant avenue of nutrient supply to developing fruits, the view has frequently been expressed that there is a substantial direct contribution to the fruit from xylem sap. This has been proposed for legumes, especially lupins (Pate, 1976b, 1980; Pate et al., 1978, 1979b) and cowpea (Pate et al., 1985), as well as for castor bean (Hocking, 1982), *Yucca flaccida* (Van Die and Willemse, 1980) and the Australian native species *Grevillea leucopteris* (Hocking, 1983b). The volumes of xylem sap acquired by these developing fruits are often stated to exceed the volumes of phloem sap required, and it is frequently claimed that the xylem transports major proportions of total calcium, magnesium and phosphorus into these fruits, as well as most of the water.

There are, however, several considerations that seriously undermine the validity of nutritional budgets for any fruit based on the composition of xylem sap and phloem sap and their arbitrary proportional representation made to fit final fruit or seed composition. First, the use of vacuum extraction to obtain samples of xylem sap close to the point of fruit attachment cannot indicate the actual composition of mobile xylem sap. Bollard, who developed this technique, cautions that the composition of xylem sap extracted under vacuum is not identical with that of ascending sap (Bollard, 1960, 1970). Normally, nitrogenous solutes and other cations will be adsorbed to the walls of xylem vessels. Although these solutes are available for transfer to adjacent cells including phloem (Section 4.4), they are not freely available to xylem sap moving slowly through the vessels *in vivo*.

Secondly, the assumption that phloem sap is delivered *per se* to the fruit walls or seedcoats cannot possibly be correct. Mass flow does not imply mass delivery. Unloading of solutes from sieve elements to receptive cells of fruit walls or seedcoats is highly selective, and regulated by the activities of specific transporters located in the plasmalemma. Various feedback effects are likely to ensure that the rates of unloading of individual solutes will not occur in direct proportion to their respective concentrations in phloem sap.

In developing legume fruits, each solute imported in the phloem may be allocated to the seeds, retained in the vasculature of the hull (Thorne, 1979, 1980), or acquired by the tissues of the hull. The identity of the translocated solute and the stage of development of the pod are both important determinants of the distribution of each solute arriving in the fruit. In the series of experiments with fruiting pea plants already discussed (Section 4.4), [^{14}C] glutamine contributed least label to the fruit compared to [^{14}C]asparagine and [^{14}C]homoserine when the leaf subtending the pod was attached to the plant. The acquisition of label from [^{14}C]asparagine fluctuated, with a maximum proportion delivered (ca. 20%) that was double the maximum from [^{14}C]glutamine (ca. 11%). The fruit consistently derived most label from a pulse of [^{14}C]homoserine: 23-27% of administered radioactivity from 12 to 18 days after flowering (Murray, 1985; see Section 5.3.1).

These disbursements of label bear no relation to the concentrations of each solute in fruit stalk phloem sap, where asparagine is at least 60 mM, homoserine is 22 mM, and the concentration of glutamine could be very similar to that of homoserine (Table 8). The priorities exercised by the seeds compared to the pod wall in gaining the majority of the total label recovered also depend on the identity of the solute and the stage of development of the fruit, rather than on the concentration of each solute in the phloem sap entering the fruit.

A third complication not considered in current modelling procedures is that phloem sap samples taken from the veins of developing pods do not represent the nutrients actually arriving in the fruit from the parent plant. Even when the possible artifacts of the sampling method have been overcome or allowed for, it must be realized that although pod vein sap samples do include newly arrived solutes, there is also an 'osmoregulatory' component (e.g. sucrose and asparagine retained in pod veins), together with a possible component representing delivery to seeds from current photosynthetic CO_2 fixation in the pod wall (Section 5.3.1 and another component representing remobilization of nutrients previously acquired by the hull (Section 5.5). Figure 12 provides an outline of these relationships.

In the mid to late stages of development, a marked reduction in the ratio of sucrose to amino compounds in pod vein phloem sap (as in lupins, Pate [1976b] or cowpea, Peoples et al., [1985b]) can reflect the mobilization of hull nitrogen to the enclosed seeds, as well as enhanced import of nitrogen from the supply organs of the parent plant. These contributions will vary in relative importance with stage of development, species and cultivar. Most of the seeds' imported nitrogen can be derived from the hull at one particular stage approaching maturity in some genotypes (Section 5.5).

With regard to the water relations of fruits, it is a common assumption that the volumes of water acquired by developing fruits are exactly balanced by tissue water contents plus transpiration demands (Hocking et al., 1978; Pate, 1976b, 1980; Pate et al., 1978, 1979; Van Die and Willemse, 1980). It is not possible to reconcile the water uptake of a fruit with such an unrealistic assumption. The constraint imposed by this assumption requires far too large a flow of water into the fruit via xylem. Estimates as high as 73% for white lupin (Pate et al., 1978) and 82% for *Yucca flaccida* (Van Die and Willemse, 1980) have been made. This invalid assumption about the proportion of the total water gained from xylem leads in turn to erroneous conclusions about the proportions of each nutrient acquired from the phloem. For example, only 26% of phosphorus and 7% of calcium were estimated to enter *Yucca flaccida* fruits via phloem (Van Die and Willemse, 1980).

* * * * * *

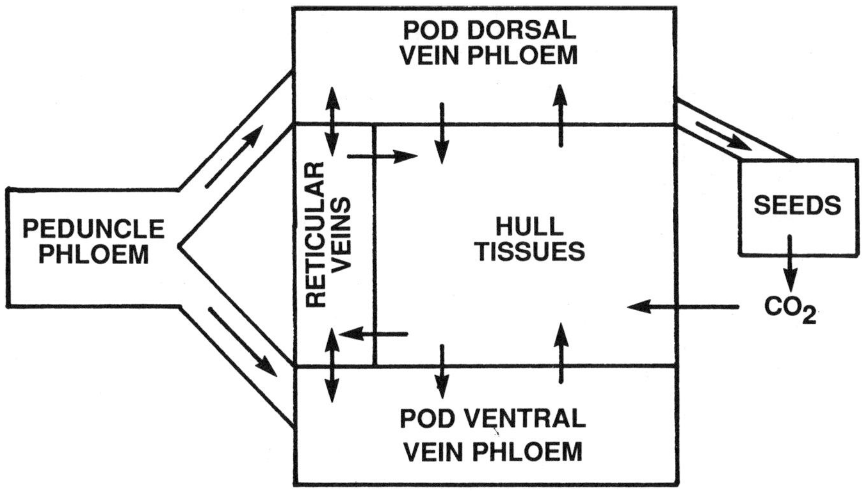

FIG. 12. Schematic diagram of the principal solute exchanges in a developing legume fruit. The transactions in any particular direction alter with stage of development and genotype, reflecting the changing abundance of transporters involved in phloem loading and unloading.

* * * * * *

Most of the water entering fruits does so via phloem; there is only a minor contribution from xylem. Recent data supporting this contention havecome from studies on tomato fruits (Ehret and Ho, 1986; Wolterbeek and de Bruin, 1986 a,b), Day-time growth rates, measured as increase in fruit diameter, can be as much as three times greater than those observed at night (Fig. 13). When fruit stalks are steam-girdled to abolish the input from phloem, this diurnal

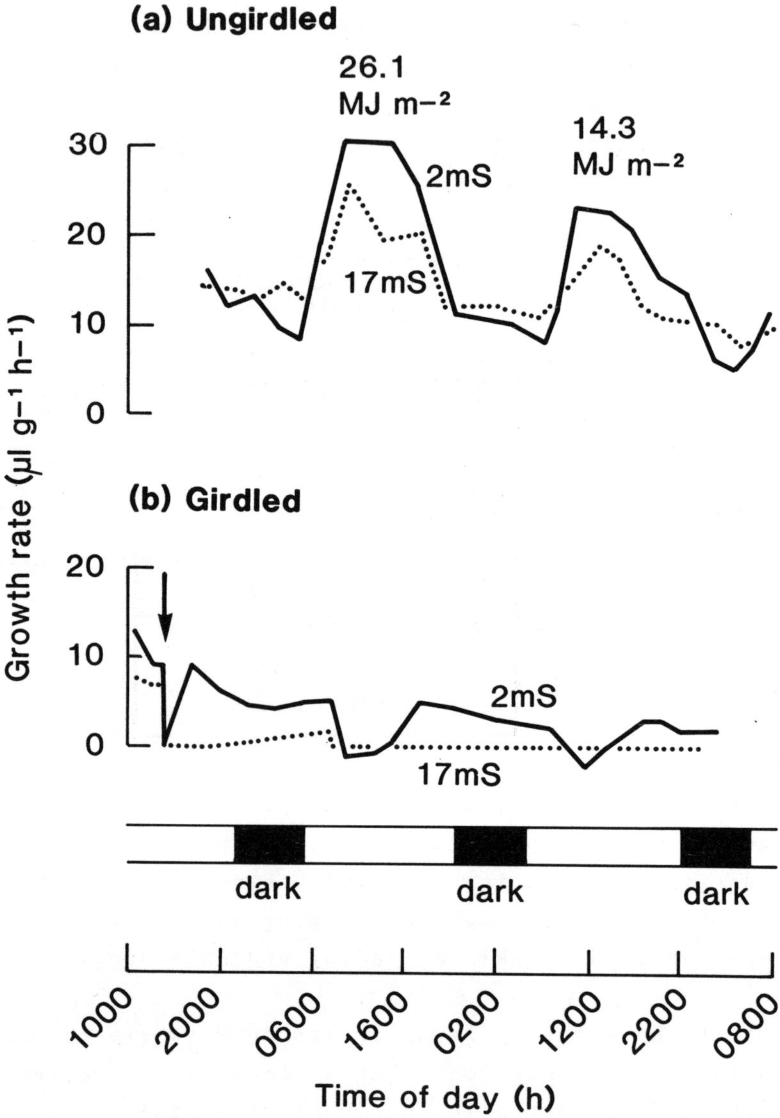

FIG. 13. Diurnal fluctuations in the rates of growth of ungirdled (a) and steam-girdled (b) tomato fruits (cv. Marathon) provided with two different external nutrient regimes, of electrical conductivity 2mS (solid line) or 17 mS (dotted). Data of Ehret and Ho (1986).

pattern reverses, with virtually no growth in the photoperiod, and a reduced rate of growth occurring at night. These observations indicate that day-time growth increments result from the net import of solutes and water from phloem. At night, the residual growth increment in the girdled fruits is produced solely by water imported from xylem. Thus about 90% of the water requirements of the developing tomato fruit are supplied by phloem, and about 10% by xylem.

The direct contribution of xylem sap to the nutrition of developing legume fruits has been considerably overestimated. Even when the delivery of surplus water through the phloem is recognized, and estimates are made of the water returned to the peduncle and stem (Section 5.2.2), wrong conclusions are drawn about the contribution of xylem sap. This is because of the first erroneous assumption about the mobility of xylem sap contents discussed earlier. In tropical legumes, the ureides allantoin and allantoic acid are abundant in xylem sap, and are major storage forms of nitrogen in the stem (Herridge et al., 1978). It is not likely that substantial quantities of ureides enter the fruits of tropical legumes via xylem. Consequently, it is unrealistic to expect that a ratio of xylem:phloem import of 9:11 (by volume) can be applicable to the first 11-day period of cowpea pod development (Peoples et al., 1985). Another impossible outcome of this arbitrary volume ratio is the conclusion that 70% of the fruit's requirement for serine must be met by synthesis within the fruit (Table II of Peoples et al., 1985a). This estimate is not only incompatible with the weak photosynthetic activity of the hull, but also with the evidence that serine obtained by the legume fruit is generated mainly in the leaves that supply the developing pods (Section 3.1.3).

5.2.2 Water Removal Through Xylem

Fruits that remain fleshy control their transpirational water losses throughout development until maturity; fruits that dry out exhibit a sharp peak in transpirational loss during final desiccation, as in pea (Flinn et al., 1977) and cowpea pods (Pate et al., 1985). In earlier literature it was frequently suggested that developing fruits take up more water from their vascular supply than actually needed for growth plus transpiration. The surplus is considered to be exported in xylem, sometimes in favour of nearby leaves under water stress (Bollard, 1960, 1970; Zimmermann 1969). The return of water to xylem is implicit in the pressure-flow model of phloem transport, and was proposed by Münch himself to operate in developing fruits (see Zimmermann, 1969).

This idea has been tested with legume pods, using labelled solutes in conjunction with marker compounds for apoplastic pathways, such as pyranin (Mix and Marschner, 1976), inulin (Simpson et al., 1983; Van Bel, 1984) and acid fuchsin.

Anatomical studies of developing pods of bean and cowpea reveal the interconnected nature of the xylem and its potential for the redistribution of water within the tissues of the hull, and the removal of water from the fruit to the veins of the pedicel, peduncle and stem (Mix and Marschner, 1976; Pate et al., 1985). Acid fuchsin supplied to the cut stem proliferates into the pod veins from the ventral strands first, with heaviest accumulation in the ventral and interconnecting veins at night (Pate et al., 1985). Conversely, acid fuchsin supplied to the cut distal tip of the pod moves gradually out of the pod veins and into the peduncle, and in some instances into the subtending leaf.

These observations support the view that there is a diurnal fluctuation in water removal from the fruit by xylem with maximal back-flow late in the photoperiod, while the uptake of water from phloem sap is greater than the immediate requirements of cells and day-time transpiration. Some of the water withdrawn from the pod via xylem contributes immediately to the transpiration losses of the fruit stalks, which are greater by day than those of the fruit itself.

Specialized functions for the different xylem strands of the pod have been tentatively identified. That the ventral strands might carry water either in or out is suggested by their preferential staining with acid fuchsin. Moreover, if it is assumed that nitrate is carried to cowpea fruits entirely in xylem, then only the xylem of the ventral marginal veins appears capable of releasing nitrate to the hull (Atkins et al., 1986). The presumption of nitrate entry from ventral xylem in cowpea pods is based on the observed induction of nitrate reductase activity in the tissues immediately surrounding the ventral veins, and the adjacent regions of the hull; the tissue surrounding the dorsal veins exhibits only 2% of the total nitrate reductase activity of the fruit. By comparison, asymmetric induction of nitrate reductase activity is not observed in the hull of developing lupin pods (Atkins et al., 1979).

It seems likely that the xylem of the dorsal strands is specifically responsible for receiving water expelled from seeds connected alternately to some of the dorsal marginal veins (Mix and Marschner, 1976; Pate et al., 1985; Section 7.6). The xylem of the ventral strands may carry water in either direction, with fluxes reversing diurnally. Whether other fruits of legumes behave like the pods of *Phaseolus vulgaris* cv. Saxa and cowpea cv. Vita 3 is a matter for further observation. The size of the fruit may have some bearing on how the xylem functions, since types of cowpea with smaller pods than those of cv. Vita 3 do not behave in the same way (Atkins et al., 1986).

5.3 Effects of Illumination on Fruit and Seed Development

5.3.1 The Photosynthetic Capacity of Green Fruit Tissues

Almost all fruits are green initially, as petals or other floral parts cannot totally prevent light penetrating responsive cells of the carpel or accessory fruit walls. Chloroplast development in fruit walls, seedcoats or embryos can be blocked in certain genotypes, as established by Mendel for pea (see Peters, 1959). How effective the green parts of fruits might be in conducting photosynthesis is a question often posed in studies of the overall economy of fruit and seed development. If green fruits are capable of fixing CO_2 to any appreciable extent, is it atmospheric CO_2 or CO_2 generated by respiration internally that is mainly fixed? Are the products of photosynthetic CO_2 fixation retained in the fruit walls, or are they preferentially reallocated to seeds?

The photosynthetic capacities of green oranges, lemons, avocados and cucumbers were assessed by Bean and Todd (1960), Todd et al. (1961) and Bean et al. (1963a,b). Rates of $^{14}CO_2$ fixation expressed per unit surface area eventually declined with increasing size of the fruit, and with decreasing chlorophyll contents (Table 13). For all these fruits, the maximal photosynthetic rates did no more than compensate for respiratory and photorespiratory losses of CO_2 while the intact fruits were illuminated. There were no positive gains from atmospheric CO_2 that might additionally compensate for respiratory losses by night. Even green fruits displaying maximal photosynthetic capacities (Table 13) would have been in negative carbon balance over each full day, except for the import of sucrose and other nutrients from the phloem.

The stable products of $^{14}CO_2$ fixation in young lemons and oranges were mainly sucrose, with significant label

Table 13.

Photosynthetic and Respiratory CO_2 Exchanges of Intact Attached Orange and Cucumber Fruits Illuminated at 20°C

Fruit fresh weight (g)	Surface area (cm^2)	CO_2 fixation rate[a]	Ratio CO_2 fixed/lost
Orange:			
43.14	68.4	2.50	1.04
62.05	98.5	0.66	0.31
Cucumber:			
2.0	7.0	2.00	0.71
6.0	18.0	1.99	0.73
22.0	42.0	1.66	0.56
26.2	47.0	2.40	1.00
61.0	93.0	4.53	0.63
67.0	101.0	5.49	0.65
90.9	127.0	1.86	0.59
124.8	160.0	1.95	0.61

[a] mg CO_2 h^{-1} per 100 cm^2. Data of Todd et al. (1961).

* * * * * *

also in glucose, alanine, serine, glutamate and organic acids (Bean et al., 1963b). This spectrum of products very clearly implicates the chloroplasts, with their ribulose bisphosphate carboxylase - oxygenase activities, as chiefly responsible for the measured CO_2 fixation (Chapter 3). The activities of PEP carboxylase reported for fruit and seed tissues are never fully expressed in terms of effective CO_2 fixation, and the function of PEP carboxylase in this context must be sought in the regulation of pH and other aspects of metabolism (Davies, 1973, 1986; Kennedy, 1986).

Most studies on the photosynthetic capacity of legume hulls have been performed with pea. Lovell and Lovell (1970) showed that closely matched pods directed a similar proportion of label derived from fixed $^{14}CO_2$ to the seeds, whether the pod was attached to the parent plant, or cut off prior to feeding. This experiment demonstrates that photosynthetic products of the hull do not need to be exported from the fruit to the plant in order to be loaded into phloem directed to the seeds. Instead, products of photosynthetic CO_2 fixation in the hull are taken up into phloem within the hull itself (Fig. 12). Other studies have confirmed that the seeds themselves do not directly fix more than trace quantities of $^{14}CO_2$ (Bhardwaj and Raju 1972; Flinn et al., 1977).

The proportion of labelled assimilates that is retained in the pea hull varies with stage of development. Young hulls fix more $^{14}CO_2$ and retain most of the products within the hull; older hulls fix less, but direct a higher proportion of labelled assimilates to the seeds (Lovell and Lovell, 1970; Bhardwaj and Raju, 1972). The period of greatest capacity to fix CO_2 corresponds to the stages around maximal chlorophyll and nitrogen contents, well in advance of maximal fresh weight and dry matter contents, but after the phase of rapid pod extension (Fig. 14).

An opposite trend to that described for pea has been observed in chick pea. At the earliest stage studied (9 day) developing pods exported about half the label derived from $^{14}CO_2$ fixation to the seeds. This proportion declined steadily to about 10% at 27 days (Singh and Pandey, 1980). In an experiment with plants trimmed in various ways, the pods of totally defoliated plants could fix only about 5% of the $^{14}CO_2$ fixed by intact plants. Singh and Pandey concluded that the pod wall contributed to its own growth, but was not a significant source of photosynthetic assimilate for seed development.

Pea pods appear to be more effective in fixing CO_2 from both the external and internal atmospheres than pods of chick pea or cowpea. However, the pea hull is incapable of significant CO_2 fixation from $^{14}CO_2$ injected into the gas cavity in darkness (Flinn et al., 1977). This inability is reflected in the very high CO_2 concentrations typical of the internal atmosphere in pods: up to 4.3% (v/v) in pea (Harvey et al., 1976) and 2.2% in cowpea (Peoples et al., 1985b). Even in pea, there is no stage when the gain of CO_2 by fixation in the light compensates for the loss of CO_2 in respiration overnight (c.f. Table 13).

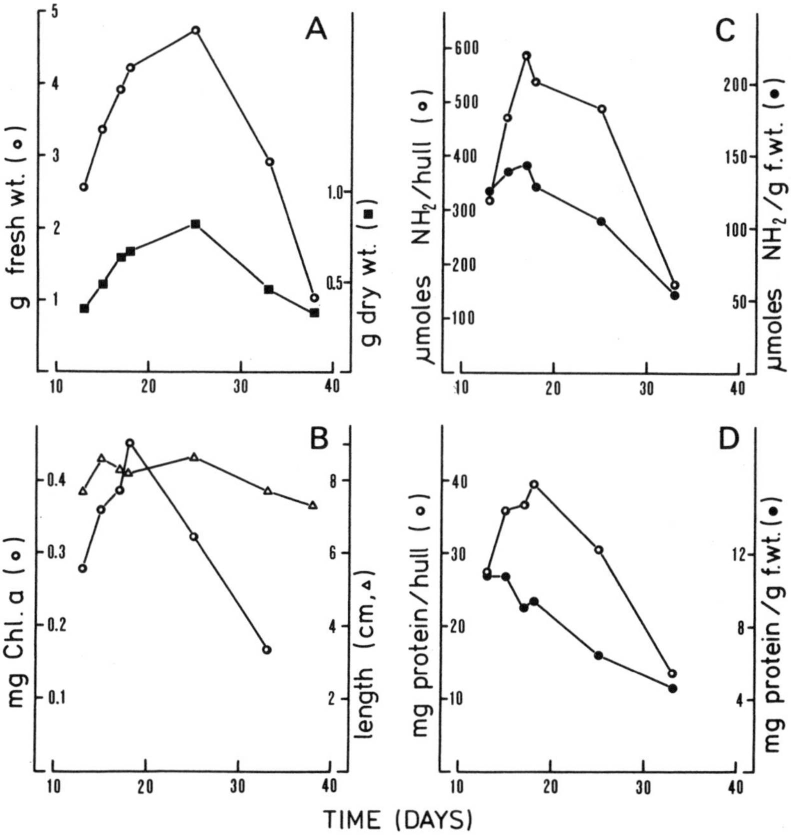

FIG. 14. Changes in: A, fresh weight and dry matter contents; B, length and chlorophyll *a* content; C, free amino nitrogen; D, extractable protein content of the hull of a developing pea pod. From Murray (1983b).

Nevertheless, Flinn et al. (1977) estimated that a pea pod subjected to diurnal illumination required 19% less carbon as assimilates from the parent plant than an equivalent pod kept in continuous darkness. Differences among pea genotypes (Bhardwaj and Raju, 1972) indicate that not all cultivars can produce pods performing as effectively as this. A systematic study of the effects of pod illumination on productivity needs to be undertaken.

The effects of maintaining individual pea pods in darkness (aluminium foil wrapping) are in fact most pronounced in the first two to three weeks following full blossom. Among a set of garden pea plants fed [U-^{14}C]homoserine, one was included with a 'dark-grown' pod of the same chronological age as one of the normal green stages (Table 14). In the pod where chloroplast development had been prevented, the acquisition of label from stem-fed [U-^{14}C] homoserine was less than half that of a green pod the same age or younger. Most of the label in this pod was retained in the hull, an allocation typical of a younger green pod.

Table 14.

Recovery of Radioactivity in the Pea Pod (cv. Melbourne Market) 24 h After Pulse-Feeding [U-^{14}C]Homoserine via the Cut Stem.

Age of pod (days)	% of fed ^{14}C in fruit	Distribution of label: % in hull	% in seeds
12 - 13	23.7	87.0	13.0
16 - 17	26.7	38.1	61.9
18 - 19	26.8	43.5	56.5
21 - 22	16.8	42.6	57.4
16 - 17[a]	10.7	73.8	26.2

[a] pod enclosed with aluminium foil throughout development

Data from Murray (1985).

* * * * * *

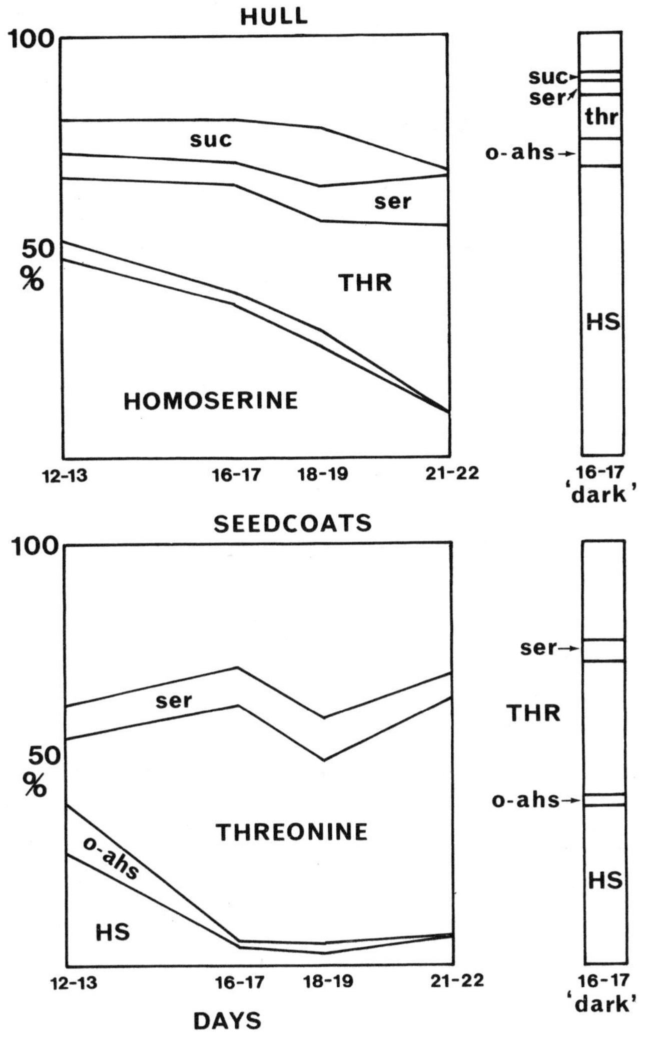

FIG. 15. Distribution of label as % of total in ethanol-soluble metabolites of the hull and seedcoats (above) and the embryos (next page) from individual fruits fed [U-^{14}C]homoserine via the cut stem at different stages of development. Cultivar: Melbourne Market. Abbreviations: standard except HS (homoserine), o-ahs, *O*-acetylhomoserine.

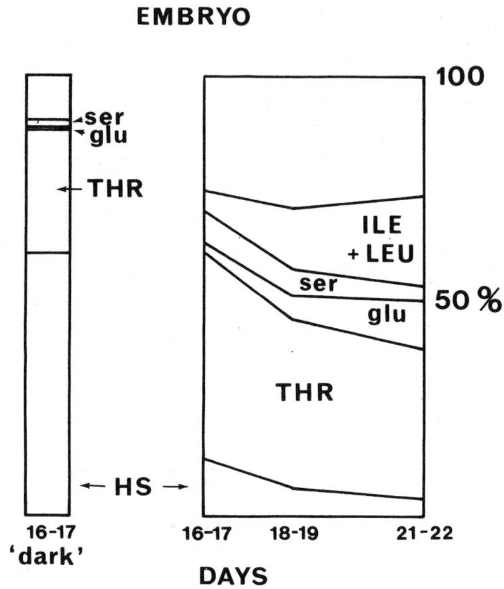

FIG. 15. Embryo portion. See preceding caption. Data from Murray (1985).

* * * * * *

The seeds from the pod deprived of illumination weighed 40% less than green seeds of the same age; within these seeds the embryos were approximately only 20% of the fresh weight of the green embryos. The conversion of homoserine to threonine, its immediate metabolic derivative, was not prevented in the seeds of the etiolated pod (Fig. 15), but the rates of metabolism in the hull, seedcoats and embryo were clearly retarded, in keeping with their slower growth rates.

5.3.2 Light Dependence of Fruit Development

Exposure to light is necessary for the normal development of fruits of most species (exceptions being those with subterranean fruits), irrespective of the contribution of the chloroplasts to the carbon and nitrogen economies of

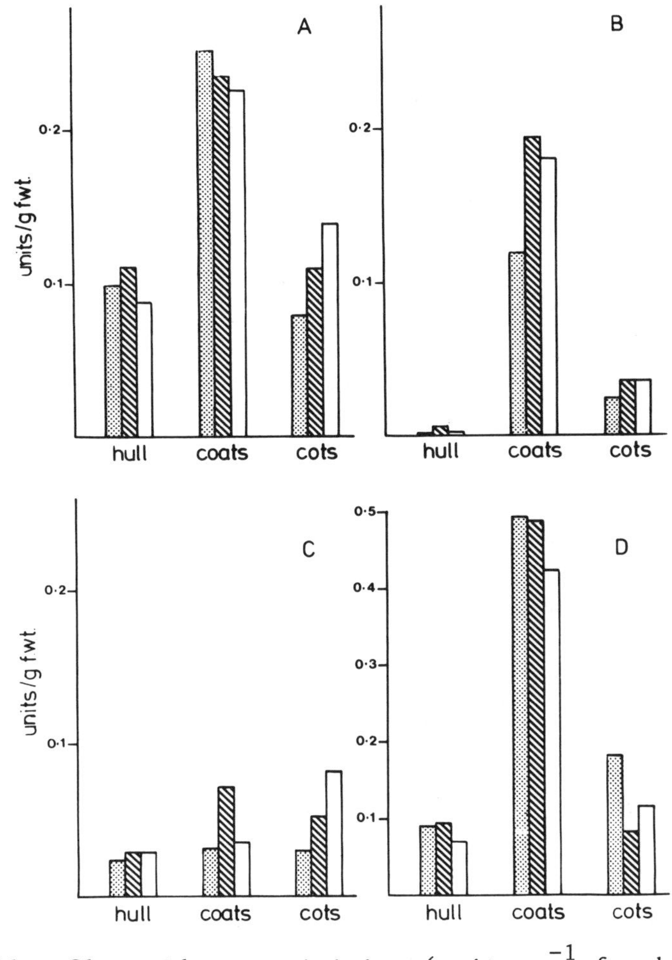

FIG. 16. Glycosidase activities (units g^{-1} fresh weight) in the hull, seedcoats and cotyledons (cots) from pea pods developing in darkness (two examples, stipled and hatched) or under normal conditions (unshaded). A, α-mannosidase; B, β-glucosidase ; C, α-galactosidase; D, β-galactosidase - note the scale change for D. Each pod was sampled at or close to maximum fresh weight. From McGee & Murray (1986).

* * * * * *

fruits and seeds. The hulls of etiolated pea pods lack chloroplasts (and consequently protein), but in length and activities of predominantly vacuolar glycosidases (McGee and Murray, 1985, 1986), they develop in similar fashion to a green pod (Fig. 16). However, the growth retarding effect of lack of illumination is so great, especially for the seeds, that effects other than the inability to fix CO_2 must also be sought.

The likely significance of local photosynthesis in the day-time production of O_2 has been suggested (Murray, 19-87). A restricted supply of O_2 would hinder rates of respiration supporting ATP-dependent phloem unloading, and aspects of carbon and nitrogen metabolism depending on mitochondria (Section 7.3.1). The possibility that the transport proteins involved in the unloading of phloem depend on phytochrome-mediated events for their synthesis and co-ordinated functioning should also be considered and investigated.

Light is necessary for the synthesis in fruit walls of accessory photosynthetic pigments (β-carotenes and oxycarotenoids) and also those unrelated to photosynthesis, but responsible for the characteristic colour changes observed in ripening fruits. In apples, the synthesis of anthocyanins concentrated in the skin is both light-dependent, and responsive to a diurnal temperature cycle (Tan, 1980). Anthocyanin production depends on the activity of phenylalanine:ammonia lyase, which in plants is regulated by phytochrome. The light-dependent synthesis of anthocyanins in apples can be modified by nutritional imbalances; nitrogen and potassium deficiencies lead to increased PAL activity and increased pigment production (Tan, 1980).

5.4 Accumulation of Sugars in Fruits

Sucrose is the main carbon and energy source translocated to fruits, entirely in the phloem. Species differ considerably in the extent to which the sucrose imported by fruit tissues is retained unmodified, or metabolized so that free hexose sugars accumulate instead. Examples of fruits that retain most of their accumulated sugar as sucrose are pineapple, peach and apricot (Bollard, 1970; Wills et al., 1983). Most cultivated legumes store sugars mainly as free hexoses (e.g. *Phaseolus vulgaris*, Sacher, 1966; soybean, Thorne, 1979). Pea pods behave like this at first, but then retain an increasing proportion of total sugar as sucrose (Bisson and Jones, 1932). Among stone fruits, considerable differences in absolute amounts of sucrose, glucose and fructose appear to be typical of different cultivars (Wills et al., 1983).

It is even possible for different regions of the one fruit to accumulate sugars differently. The green and orange regions of the cantaloupe (or rock melon, or netted muskmelon; *Cucumis melo* var. *reticulatus*) both accumulate hexoses until after 20 days from fertilization. Then the sucrose content increases dramatically, from below detectable amounts up to contents matching the hexoses, especially in the orange tissue (Lingle and Dunlap, 1987). The increase in sucrose content is correlated with declines in acid and neutral invertase activities, together with an increase in sucrose phosphate synthase, an enzyme necessary for sucrose synthesis (Section 3.2.1). This implies that most imported sucrose is no longer hydrolysed following import, and that hexose pools contribute also to the increasing sucrose content.

Acid invertase is responsible for the hydrolysis of sucrose in flower buds (Russel and Morris, 1978) and fruits such as apple and grape (Hawker, 1969, 1985). Glucose and fructose accumulate in fairly equal proportions in grapes, up to very high contents (about 20% w/v in expressed juice [Coombe, 1960; Hawker, 1985]). It seems more probable that the hydrolysis of sucrose by acid invertase should occur in the vacuole (Hawker et al., 1976) rather than as an obligatory part of unloading of sucrose from phloem.

Alternative mechanisms for dismantling sucrose also

occur, particularly in seedcoats and embryos, which lack acid invertase activity (Section 6.3.3). In such instances the sucrose synthase reaction is considered to work in reverse (Matheson, 1984; Hawker, 1985; Edwards and ap Rees, 1986a,b):

$$\text{UDP} + \text{sucrose} \longrightarrow \text{UDP-glucose} + \text{fructose}$$

The subsequent metabolism of UDP-glucose involves donation of glucose units in the synthesis of polysaccharides, and conversion to glucose-1-phosphate by the action of UDP-glucose pyrophosphorylase (Turner et al., 1957; Huber and Akazawa, 1986):

$$\text{UDP-glucose} + \text{PPi} \longrightarrow \text{UTP} + \text{glucose-1-phosphate}$$

Both fructose and glucose-1-phosphate derived from these two reactions could then contribute to hexose phosphate pools involved in glycolysis. Multiple forms of phosphoglucoisomerase and phosphoglucomutase are found in hulls, seedcoats, cotyledons and embryonic axes from developing pea pods (Murray and Ayre, 1987).

5.5 Redistribution of Nutrients from Fruit Tissues to Seeds

Bollard (1970) drew attention to a lack of information on "the competition for nutrients between developing fruit parts and regulation of their movement to different tissues in the developing fruit". It is well known that the distribution of most mineral elements within a fruit is not homogeneous, but how this eventuates is not understood very well. For example, the apical region of a sound tomato fruit contains more potassium and nitrogen, but less calcium and magnesium than the region adjacent to the stalk (Marcelle and Bodson, 1979). Heterogeneity of cation deposition with phytate has been monitored at a cellular level in seed storage tissues (Lott, 1984; Mazzolini et al., 1985) but similar information for fruit tissues is technically more difficult to obtain and is not available.

Remobilization of nutrients from fruit walls in favour of seeds has been studied for a number of legumes, castor bean, *Grevillea leucopteris* and *Xanthium occidentale* (Hocking, 1982, 1983a,b). The time-course for net accumulation and mobilization varies for individual nutrients, and even among legumes, the extent of apparent remobilization can differ remarkably (Hocking and Pate, 1977).

In pea pods, early gains to maximum contents are observed for nitrogen (Fig. 14), P, Ca, Mg, Zn, Fe, Mn and Cu, whereas the gain in K more closely matches the increase to maximum dry matter content (Hocking and Pate, 1977). For most elements, remobilization commences before the maximum dry matter content is attained, and ultimately about 40% of this maximum dry matter content is retained (Fig. 14). Most of the carbon exported from the hull to the seeds must first be reconverted to sucrose. Thorne (1979, 1980) considers that sucrose is not found in significant quantities in the soybean hull outside the phloem. However, so far there have not been any comprehensive studies of the enzymes involved in this process as there have for fleshy fruits (Section 5.4).

More than 80% of the maximum nitrogen and phosphorus contents of pea hulls are mobilized, with values for apparent removal ranging between 27% to 48% for the other elements studied, except Ca and Na (Hocking and Pate, 1977). The allocations of Ca and Na to the pod walls of pea and lupin are virtually permanent, with at most about 10% remobilized to seeds. The hull of white lupin behaves in a similar fashion to the pea hull, but apparent recoveries of nutrients from the hull of blue lupin tend to be lower.

Although the contribution from the pod wall often appears to be a minor proportion of the total nutrient acquired by the embryo, in the period that their greatest contributions are made, imports from the hull can dominate a seed's current import. This has already been emphasized in the case of nitrogen, where the net removal of amino nitrogen from the pea hull (cv. Melbourne Market) contributed an estimated 28% of the amino nitrogen stored in similar forms in the mature seeds (Murray, 1983b). However, the net removal of amino nitrogen from the hull after 25 days and up until 33 days accounted completely for the increment in amino nitrogen (free plus protein-bound) in the developing embryos during this period of reserve protein synthesis.

The treatment of the hull as a dynamic reservoir by Bisson and Jones (1932) has been confirmed. The hull of garden pea contains a wide spectrum of prominent nitogenous solutes while the hull's contents of free amino nitrogen and protein are both increasing to their maxima (Fig. 14 C, D). In cv. Melbourne Market these solutes are homoserine, asparagine, glutamine, serine, alanine, valine, aspartate and glutamate. Each of these compounds shows a characteristic period of maximum abundance. Even the two predominant solutes with the appearance of most stability show evidence of significant turnover in pulse-labelling studies (Murray and Cordova-Edwards, 1984a,b; Murray, 1985, 1986a).

The mobilization of protein from the hull involves the concerted action of a range of proteolytic enzymes (Storey and Beevers, 1977) and the interconversion of the resultant free amino acids to forms preferred for translocation to the seeds. These processes include the removal of chloroplast proteins and consequent dismantling of the hull's limited photosynthetic capacity. Once the hull has made its major contribution to seed nitrogen import, the final stages of protein accumulation in the cotyledons must depend almost entirely on supplies of assimilated nitrogen mobilized from the rest of the plant.

Some contrasting patterns of nutrient redistribution for

dry fruit types other than legumes have been described by Hocking. Dehydrating capsular tissues of castor bean mobilize Zn substantially (73% withdrawn), phosphorus (42%) and Cu (23%) moderately, and Mn slightly (8%). They retain practically all their N, K, S, Ca, Mg, Na and Fe, and show no net loss of dry matter (Hocking, 1982). In follicles of *Grevillea leucopteris* there is also no apparent redistribution of dry matter, but at maturity, the woody follicle has mobilized almost all of its P, and a minor proportion of Zn. No apparent mobilization occurs for Mn, Mg, Cu, N, Ca, Fe, S, K or Na (Hocking, 1983b). Separate time-course studies were not obtained for fruits of Noogoora burr (depicted in Murray 1986c), but at maturity, the fruit walls contain little N or P; major proportions of the fruit's total K, Ca, Cl, Na and Mn are retained in the fruit walls (Hocking, 1983a).

In all of these fruit types studied by Hocking, several macronutrients commonly regarded as mobile have been immobilized in the fruit walls: N, S and Mg in two species each, and K in all three. This may be significant in relation to the ecology or management of each species, but wide-ranging conclusions cannot be drawn from the behaviour of so few species.

CHAPTER 6

The Nutritive Function of Seedcoats

The possession of two seedcoats, an outer testa and an inner tegmen, is the most common condition among angiosperm ovules. Often the testa develops specially thickened and pigmented cells, which contribute to the strength of the mature coats (Tran and Cavanagh, 1984; Murray, 1984a). Sometimes the tegmen differentiates to assume a protective function instead of, or in addition to, the testa (Eames and MacDaniels, 1947; Esau, 1960). The seedcoats may also bear outgrowths or enlarged ovule stalks (funicles) which may be attractive to vertebrate and invertebrate agents of seed dispersal (O'Dowd and Gill, 1986; Howe, 1986).

These protective and dispersal functions are well documented. Other proposed functions for the tegmen include the storage of limited quantities of reserves, and the supply of nutrients to the embryo sac throughout the course of seed development. Usually the reserves are mobilized from most cell layers of the tegmen as the seed matures, but sometimes a storage and future regulatory function is maintained. The aleurone layer of the mustard seed (*Sinapis alba*) originates as the fourth or fifth cell layer of the tegmen (Bergfeld and Schopfer, 1986). The aleurone layers of other dicotyledons may develop from the tegmen in a similar fashion. That the developing seedcoats are solely responsible for the provision of nutrients to the developing embryo is a view that has not been universally accepted. Evidence in favour of this concept is discussed in this chapter and in Chapter 7.

6.1 Alternative Hypotheses on Transfer of Nutrients to the Embryo

Exclusive roles in the nutrition of the embryo proper have been attributed by anatomists to various structures in the embryo sac, including: one or both synergids (see Newcomb, 1973); the suspensor; cellular or nuclear endosperm; and 'liquid endosperm'. Many of these views are extremely fanciful, and persistent in the literature (Raghavan, 1977; Johri, 1984).

6.1.1 The Suspensor

After the first mitotic division of the zygote, the embryo differentiates into the embryo proper and the suspensor. Before the embryo grows substantially, the suspensor first elongates, reaching a length often measured in mm. By this extension, it brings the embryo proper into a precise location within the embryo sac. Orientation of the embryo has traditionally been perceived as a major function of the suspensor, a function embodied in the nomenclature.

However, an exclusive role for the suspensor in the transport of nutrients to the embryo proper is commonly put forward (Newcomb, 1973; Nagl, 1976a,b). Suspensors are also considered to be a source of growth substances for the embryo (Przybyllock and Nagl, 1977; Lorenzi et al., 1978; Yeung and Sussex, 1979).

The small size of the suspensor has been a deterrent to physiological studies. Nevertheless, Bhalla et al. (1979, 1980, 1981a,b) and Singh et al. (1979, 1980a,b) have provided some interesting information on the changing enzyme activities and metabolic capacities of the different parts of the complex suspensor of nasturtium (*Tropaeolum majus*), the same species studied by Nagl (1976a,b). The suspensor of nasturtium has two extra arms ('haustoria') connecting the basal suspensor to the chalazal and micropylar regions of the surrounding seedcoats. Often it was necessary to dissect 50 suspensors at a given stage of development to provide enough material for extraction and assay.

The young suspensor relies predominantly on the pentose phosphate pathway, and also develops PEP carboxylase activity. It is possible to correlate many of the other enzyme activities studied by this group with the progress of differentiation in the suspensor, and its eventual disintegration. However, the data provided by these authors do not discriminate either for or against a transport function for the suspensor. Their observations are consistent with whatever functions may otherwise be determined, such as "to anchor the embryo and position it in a nutritionally favourable environment" (Bhalla et al., 1981).

Newcomb (1973) questioned whether the suspensor could provide the nutrients required for the development of the endosperm in sunflower (*Helianthus annuus*), while assuming that it could function in nutrient transfer to the embryo early in its development. Hawker and Buttrose (1980) concluded that it would not be possible for a suspensor to translocate solutes at rates high enough to support the growth of developing almond embryos (*Prunus dulcis*). Selecting the interval 20 to 24 weeks after anthesis, they calculated that assimilates necessary to account for the increase in lipid plus protein in the embryo would need to be transferred through the suspensor at a specific mass transfer rate of 15 mg s^{-1} cm^{-2}, an estimate at least 20-fold higher than values reported for phloem sieve elements (Kallarackal and Milburn, 1984, 1985b). The structural and ultrastructural features of suspensor cells do not suggest that they can function in a superior manner to phloem sieve elements.

6.1.2 Endosperm

The different patterns of endosperm development in angiosperms can be classified in two ways (Murray, 1984a). The first is the classic subdivision according to the pattern of cell wall formation expressed by endosperm protoplasts. In the fully cellular type, cell walls are established around each protoplast as it completes mitotic division. In free nuclear endosperm, the formation of cell walls is postponed, perhaps indefinitely, for most endosperm protoplasts. In the helobial type, cell wall formation is retarded after the initial division of the triple fusion nucleus, which produces a wall dividing the chalaza from the micropylar region of the embryo sac (Brink and Cooper, 1947).

The second classification comprises the following types:

(i) endosperm is formed and retained as a reserve organ in the mature seed;

(ii) endosperm is formed, but substantially degraded by the time the embryo is mature;

(iii) endosperm nuclear divisions are terminated early.

Each of these functional subdivisions needs qualification. Thus small regions of endosperm can be selectively degraded as the embryo completes its development, but the endosperm is nevertheless retained as the major reserve organ for some species in category (i). A good example is rye (*Secale cereale*), where endosperm is depleted adjacent to the scutellum prior to grain maturity (Nutman, 1939). In the seeds of some orchids, there are evidently no divisions of the endosperm nucleus at all (Eames and MacDaniels 1947). Such species remain in category (iii) - it is a matter of degree how early endosperm nuclear divisions are terminated.

My main purpose in providing this alternative classification was to clarify what was meant by 'liquid endosperm' (Section 6.1.3), to enter the debate about 'advanced' and 'primitive' characteristics (Sporne, 1954; Swamy and Ganapathy, 1959), and to relate responsibility for the provision of nutrients to the embryo sac entirely to seedcoats.

The degradation, crushing and removal of endosperm cells during seed development will provide incidental supplies of nutrients for the developing embryo. But the coincidence of embryos failing to thrive because the endosperm has failed to develop in the same ovule should not be accepted as evidence that the embryo depends upon the endosperm as a conduit for its nutrient supplies (Raghavan, 1977). Clearly the failure of both entities to thrive may have a common cause, as in the transport mutants of barley (Thorne, 1985). Similarly, developmental patterns involving endosperm breakdown should never be taken to imply that all the nutrients acquired by an embryo are coming to it via the endosperm.

In the almond, the perisperm (derived from the nucellus) develops first as a reserve tissue. Then the endosperm develops while the perisperm senesces (Hawker and Buttrose, 1980). Finally, the embryo itself provides the main repository for reserves in its cotyledons, and the endosperm deteriorates. Hawker and Buttrose (1980) proposed independently that only "the inner layers of the seedcoats" could be responsible for the transport of metabolites to the embryo throughout all stages of development (a period of 32 weeks in this species).

6.1.3 The Status of Embryo Sac Liquid

The central embryo sac liquid of pea seeds arises as a secretion from the inner seedcoat, and is distinct from the vestigial endosperm (Murray, 1979b, 1980a). Support for this interpretation stems from the observations of Cooper (1938), Marinos (1970) and Ryczkowski (1960-1969). Like Cooper, Marinos studied seed development in pea, but with the aid of the electron microscope. Marinos (1970) queried the increasing use of the term 'liquid endosperm', and showed by the careful introduction of fixative into the embryo sac that the free nuclear endosperm lines the sac in a thin layer, readily distinguished from the central vacuole. The endosperm protoplasts are evenly distributed around the central vacuole, and also surround the tiny embryo, where they become fully cellular by elaborating cell walls.

Thus Marinos confirmed the observations of Cooper (1938) and concluded that "the endospermic cytoplasm is engaged in considerable and precisely controlled wall-building activities". It is these cross-linking walls that secure the developing embryo in its invariable and characteristic position against the inner integument, allowing the expanding embryo to occupy the chalazal end of the embryo sac first (for illustration, see Murray, 1980a, 1984a, 1987). The location of the embryo depends on the growth of the suspensor (Cooper, 1938; Section 6.1.1), and differs with species. In developing chick pea seeds, the embryo first occupies the opposite end of the sac to that typical of pea, and later expands towards the chalazal end (Vairinhos and Murray, 1982).

The work of M. Ryczkowski (1960-1969) is rarely cited outside Poland (see Smith, 1973; Yeung and Brown, 1982). In studies of developing seeds from a variety of species, Ryczkowski correctly distinguished the endosperm from the central vacuole of the embryo sac, and described developmental changes in both entities. He was able to demonstrate a programmed increase, then decrease, in the osmotic value of the total solutes in the embryo sac liquid. For example, in *Haemanthus katherinae* the central vacuole sap showed an increase from 0.15 M to 0.36 M over the first 15 days following perianth wilting, after which a decline to 0.23 M took place in older ovules (Ryczkowski, 1969). The maxima recorded for the osmotic value attained by the vac-

Table 15.
Maximum Values for the Osmotic Contents of Embryo Sac Vacuolar Liquid Determined by M. Ryczkowski.

Species	Value (M)	Endosperm type	Reference
Monocotyledons:			
Haemanthus katherinae	0.36	helobial (i)	1969
Haemanthus albiflos	0.39	helobial (i)	1960a
Clivia sp.	0.46	nuclear (i)	1960a
Leucojum aestivum	0.614	nuclear (i)	1960a
Asparagus officinalis	0.311	nuclear (i)	1960a
Dicotyledons:			
Crambe tatarica	0.545	nuclear (i)	1960a
Aesculus glabra	0.46	nuclear (iii)	1969

uolar liquid from developing seeds of seven species are listed in Table 15.

Ryczkowski (1964a) considered that the increases in solute concentration in embryo sac vacuolar liquid took place "by an inflow of osmotically active compounds from vegetative organs into the ovule with water". He established that the major solutes of the liquid are sugars, both non-reducing and reducing, with each type showing characteristic changes in abundance with stage of development. He showed further, for *Aesculus glabra*, that the peak amino nitrogen content of the liquid was delayed in relation to the maximum osmotic value or sugar concentration (Fig. 17).

Ryczkowski frequently emphasized the relevance of his studies to the development of media for embryo (and endosperm) culture. His observations clearly indicated that if such media are to adequately reflect the conditions that embryos are exposed to *in vivo*, they would need to be varied for different species, and should ideally be changed as development proceeded. This theme is the subject of Chapter 8.

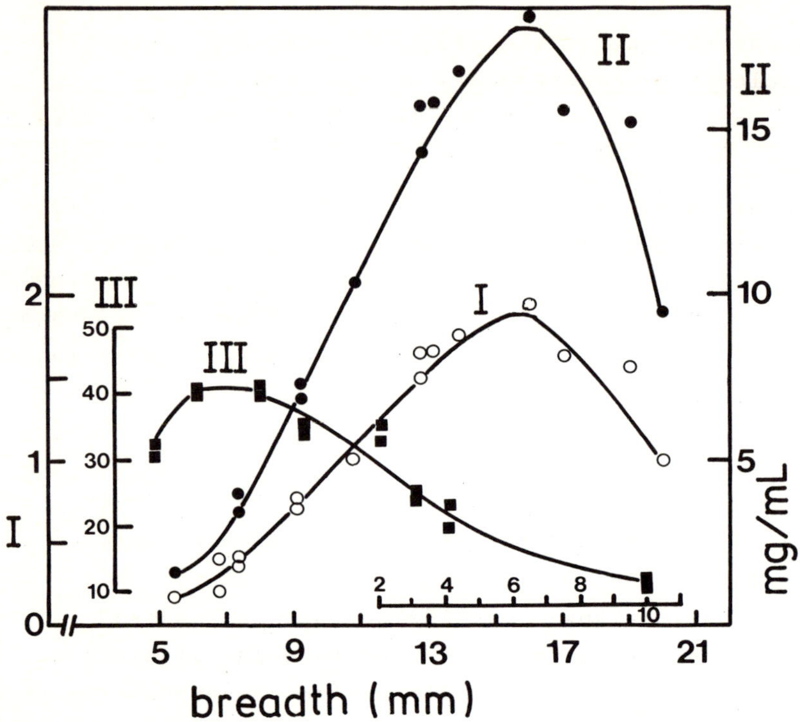

FIG. 17. Changes in solute concentrations in embryo sac liquid from developing seeds of *Aesculus glabra*. I, amino nitrogen; II, aspartate equivalent to amino N; III, total soluble carbohydrate (all as mg mL^{-1}). Main abscissa: breadth of ovule; inner scale: breadth of cotyledons. Data of Ryczkowski (1964b).

* * * * * *

The complete composition of embryo sac liquid has not yet been determined for any species except coconut (Baptist, 1963; Raghavan, 1977). Measurements of pH range from 5.2 to 5.4 in *Phaseolus vulgaris* (Smith, 1973) to pH 5.5 in pea (Murray, 1979b, 1980a). These values are consistent with those expected for vacuolar contents. Residual activities of enzymes that characterize vacuoles have also been detected. Acid phosphatase is the strongest of these at about 0.01 units per 50 μL of liquid (Murray, 1980a); only traces of α-mannosidase (0.003 units mL^{-1}) and β-galactosidase can be found.

Smith (1973) observed higher concentrations of reducing sugars (mainly glucose) than sucrose in embryo sac liquid from the earliest stages of bean seeds. He also identified organic acids, predominantly malate and citrate (23 mM and 5 mM respectively). The mineral cation in greatest abundance was K^+ (34.4 mM initially), followed by Mg^{2+} (4.1 mM) but Smith could not detect the presence of Ca^{2+}. However, Ca^{2+} is abundant in coconut milk, and precipitates as Ca phosphate if the pH is made alkaline (Baptist, 1963). Although the absolute concentrations were not stated, Hocking and Pate (1977) have shown that K^+, Ca^{2+} and Mg^{2+} are the three most abundant mineral cations in embryo sac liquid from white lupin, blue lupin and pea seeds, together comprising more than 80% of all mineral cations detected. Traces of Fe, Mn, Na, Cu and Zn were also present. Phosphorus is detected in embryo sac liquid from pea seeds entirely as Pi, in concentrations ranging between 2.6 and 8 mM (Table 16).

According to Smith (1973), the concentration of NH_4^+ in bean embryo sac liquid samples ranged between 80 mM and 130 mM, as determined by the Conway microdiffusion method. I obtained similar high values for pea seed embryo sac liquid, with some differences according to cultivar. Since I had used a Nessler reagent, it was suggested that the alkaline reagent might have substantially degraded glutamine to NH_4^+ and glutamate. A recent sample of embryo sac liquid from pea seeds gave 40 mM NH_4^+ with a Nessler reagent, but only 5 mM NH_4^+ with the Solórzano phenolhypochlorite reagent (Murray, 1987). The data of Table 16 have been modified accordingly, assuming for the purposes of calculation that asparagine made no substantial contribution (see Table 19).

Table 16.
The Concentrations of Several Macronutrients in Embryo Sac Liquid from Developing Pea Seeds (Three Cultivars).

Seed fresh wt. (g)	Liquid volume (μL)	Concentration (mM): Pi	Reducing sugars	α-NH_2	Ratio[b] Gln to α-NH_2
Melbourne Market:					
0.084	50	4.40	16.5	313	0.080
0.130	35	4.00	25.5	260	0.135
0.205	20	2.61	8.9	212	0.080
Telephone:					
0.102	45	3.78	76.0	254	0.480
0.150[a]	50	2.67	47.8	269	0.327
0.149[a]	50	3.30	31.1	379	0.269
'Sugar' pea:					
0.138	40	8.00	33.6	304	0.138
0.146	45	5.04	39.4	266	0.184
0.161	50	5.67	34.7	305	0.148

[a] seeds from the same pod; [b] attributing previously determined "NH_4^+" to the amide of glutamine, except for an estimated value of 5 mM throughout (see text).
Data from Murray (1980a).

* * * * * *

The liquid contents of the embryo sac have been shown to serve as a dynamic rather than static reservoir of nutrients. As solutes are removed by an expanding embryo (or developing endosperm, as in most species listed in Table 15), they are replenished by continued secretion from the seedcoats. Thus the amino nitrogen content of the embryo sac liquid of developing pea seeds remains approximately constant while the embryo is first expanding to occupy the chalazal half of the sac (Fig. 18). Later, as the volume of liquid contracts to 20 μL and less, a net depletion in Pi, reducing sugars and amino nitrogen becomes clear (cv. Melbourne Market stages, Table 16). Qualitative changes in nitrogen supply and their significance in relation to the changing metabolic capacities of seedcoats are considered in more detail in Section 6.3.

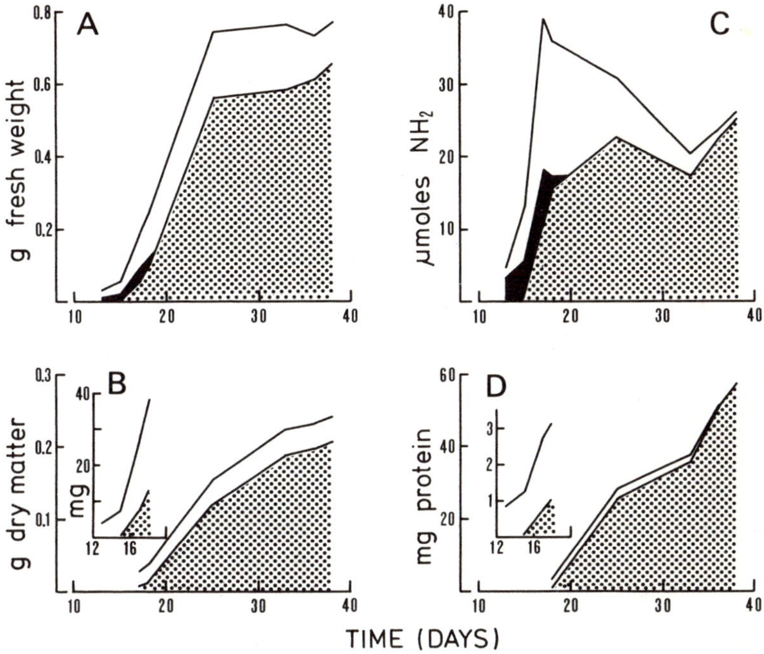

FIG. 18. Changes during development of the pea seed in: A, fresh weight; B, dry matter; C, amino nitrogen; D, protein. Parameters are expressed per seed, with seedcoats unshaded, embryo sac liquid solid (omitted from B), and the embryo stippled. From Murray (1983b).

* * * * * *

6.2 Growth Parameters and Enzyme Activities of Seedcoats

In their studies of developing pea seeds, Raacke (1957), Bain and Mercer (1966) and Flinn and Pate (1968) recorded changes in several seedcoat parameters, including dry matter content, starch, sugar and nitrogen contents. Typical changes in amino nitrogen and protein contents of pea seedcoats are included in Figure 18. Electrophoretic profiles of proteins extracted from seedcoats show a cluster of relatively low MW proteins, quite distinct from higher MW proteins typical of cotyledons (Vairinhos and Murray, 1982).

Developing seeds of pea were the first employed to provide information on changing enzyme activities in seedcoats (Table 17). The major difficulty in obtaining suitable extracts from seedcoats of species other than garden pea is their content of phenolic compounds, reflected in the association of catechol oxidase with the particulate fraction from extracts of seedcoats of the wild pea, *Pisum elatius* (Marbach and Mayer, 1974, 1975). This problem can be overcome by preparing extracts with polyethylene glycol (Hawker, 1969; Peoples et al., 1985a).

Table 17.
Enzyme Activities Assayed in Soluble Extracts from the Seedcoats of Developing Pea Seeds.

Enzyme	References
PEP carboxylase	Hedley et al., 1975
RuBP carboxylase	"
Isocitrate dehydrogenase	"
Aminopeptidases	Collier & Murray, 1977
	Murray, 1979b
Acid phosphatase	Murray and Collier, 1977
	Murray, 1979b, 1980a,b
Acid glycosidases	Murray, 1979b, 1983a
Asparaginase and enzymes)	(Sodek et al., 1980
of nitrogen metabolism)	(Murray & Kennedy, 1980

* * * * * *

A number of isoenzymes have been identified that appear to be present uniquely in seedcoats, e.g. of acid phosphatase (Figs 19, 20) and α-mannosidase (Murray, 1983a). Electrophoretic profiles for several enzyme activities not included in Table 17 show that generally the seedcoats share isoenzymes present in other organs, although often the combination and quantitative distribution of multiple forms is characteristic of seedcoats (Murray and Ayrė, 1987).

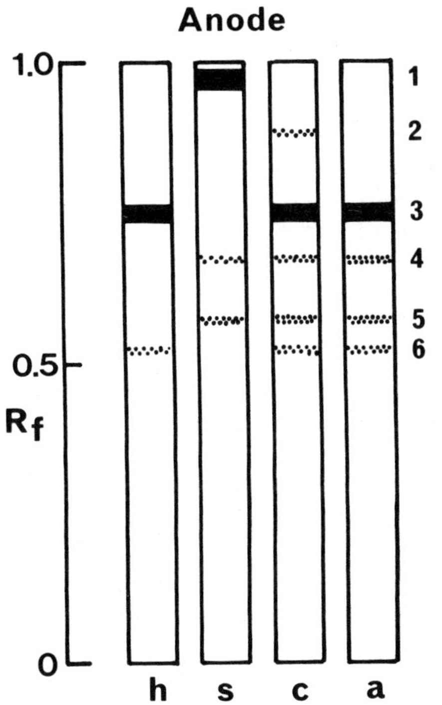

FIG. 19. The mobilities of the multiple forms of acid phosphatase detected following polyacrylamide gel electrophoresis of extracts from hulls (h), seedcoats (s), cotyledons (c) and embryonic axes (a) of developing pea fruits (cv. Telephone). From Murray and Collier (1977).

* * * * * *

The total α-mannosidase activity of the seedcoats can be taken as a marker for vacuole development (Fig. 16). The maximal activity of acid phosphatase in seedcoats occurs mid-way through development, after maximal α-mannosidase

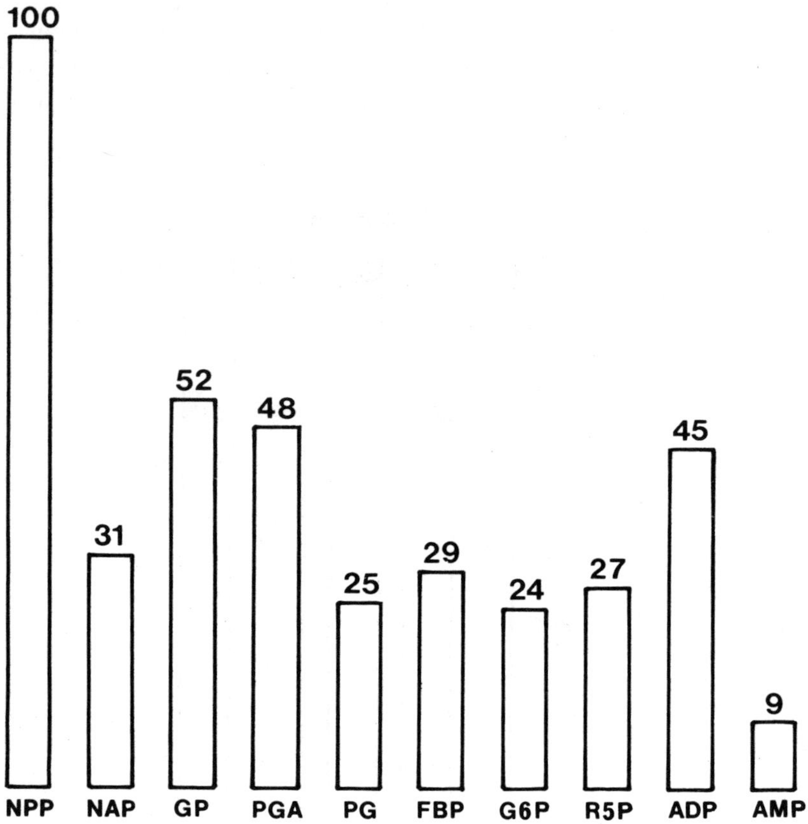

FIG. 20. Substrate preferences of an acid phosphatase isolated from the seedcoats of developing pea seeds. NPP, *p*-nitrophenylphosphate; NAP, α-naphthyl acid phosphate; GP, β-glycerophosphate; PG, phosphoglycollate. Data of Murray (1980b).

* * * * * *

activity has first been attained (Fig. 21). The changes in acid phosphatase activity parallel changes in phosphate ester content more closely than changes in Pi content, until maturity, when the only remaining P in the seedcoats is in the form of Pi. The acid phosphatase activity that persists in mature seedcoats is more than adequate to ac-

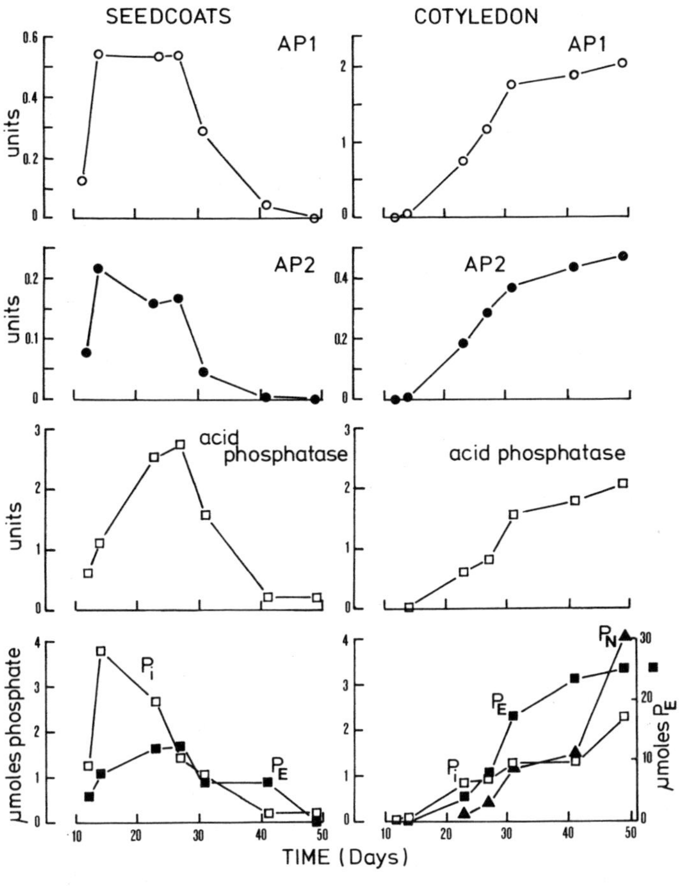

FIG. 21. Changes in the activities of aminopeptidases AP1 and AP2, acid phosphatase (enzyme units per plant part) and in the contents of Pi, P_E and P_N (nucleic acid P) in seedcoats and a single cotyledon of developing pea seeds (cv. Telephone). From Murray (1979b).

* * * * * *

complish the complete removal of phosphate esters (Fig. 21). Aminopeptidases AP1 and AP2 probably assist in the breakdown and recycling of signal peptides cleaved from newly synthesized polypeptides during periods of net synthesis of proteins (young seedcoats) as well as in the later mobilization of protein from the seedcoats.

The significance of some of the changes in enzyme activity that take place as seedcoats develop is not always clear. Profiles of changing enzyme activity need to be augmented with other ways of studying metabolism and function (Section 6.3). The significance of the changes in PEP carboxylase activity in pea seedcoats (Hedley et al., 1975) was not immediately obvious. This activity is much greater than that of RuBP carboxylase, and declines after an early maximum. The PEP carboxylase activity measured *in vitro* is not matched by any marked ability of the intact seed to fix $^{14}CO_2$ (Lovell and Lovell, 1970; Flinn et al., 1977). However, because the PEP carboxylase reaction generates a proton (see p.92 of Kennedy, 1986), and the seedcoat secretion is acid, the possibility that PEP carboxylase acts only in accordance with pH stat requirements must be considered (Section 6.3.2).

6.3 Transformation and Secretion of Metabolites by Seedcoats

That seedcoats are responsible for secreting nutrients into the embryo sac has been confirmed by experiments in which the embryo is removed and replaced with a collection medium, either agar (Fig. 22), or liquid. This technique allows the composition of the seedcoat secretion to be monitored over relatively short time periods, and at stages of development after the embryo sac liquid has been displaced.

The earliest studies with these 'evacuated' seedcoats were performed with soybean (Thorne and Rainbird, 1983; Thorne and Hardy, 1984), common bean (Patrick, 1983, 1984) broad bean and garden pea (Wolswinkel and Ammerlaan, 1983; Rainbird et al., 1984). The rates of release of amino nitrogen by seedcoats functioning in the absence of an enclosed embryo are comparable with the rates that would be

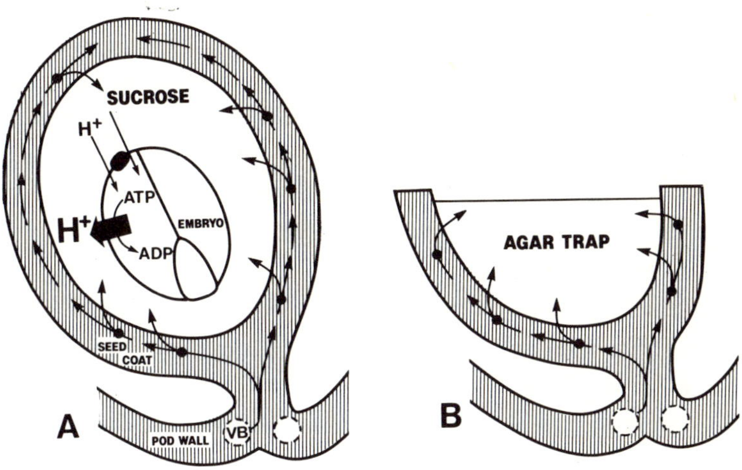

FIG. 22. Diagram of intact (A) and modified (B) soybean seeds showing possible routes of solute transfer through seedcoats. In (A), the embryo is displaced from its normal position against the tegmen. An agar trap (B) can replace the embryo and collect the solutes secreted by the seedcoats. From Thorne and Rainbird (1983).

* * * * * *

expected in intact fruits attached to the plant. For example, de Ruiter et al. (1984) measured release rates for total amino nitrogen plus NH_4^+ of between 1 and 2 μmol h^{-1} per seed. These rates are consistent with estimates of 1.1 μmol amino nitrogen assimilated into cotyledonary protein per hour in intact pea seeds (Murray, 1979b). Direct measurements of secretion rates and composition (Section 6.3.2) complement studies of changing enzyme activities *in vitro*, and pulse-chase studies with labelled solutes *in vivo*.

6.3.1 <u>Secretion of Phosphate</u>

The sole form of phosphorus accumulated in embryo sac

liquid is Pi (Table 16). Wolswinkel and Ammerlaan (1985a) compared the rates of secretion of Pi, amino nitrogen and sucrose by attached and detached evacuated seedcoats. After detachment at the funiculus, the rates of secretion of Pi are maintained for relatively long times (Fig. 23). The rates of secretion of sucrose and total amino nitrogen are curtailed much sooner. In pulse-chase studies where ^{32}Pi was supplied to soybean pods via the stem, label accumulated in the seedcoats before any secretion took place (Neumann and Noodén, 1984). Taken together, these observations suggest that a second pool of Pi, possibly vacuolar (Woodrow and Rowan, 1979), could regulate the availability of Pi for secretion. By comparison, pools of sucrose and amino compounds destined for secretion from seedcoat cells appear to lack buffering from other compartments.

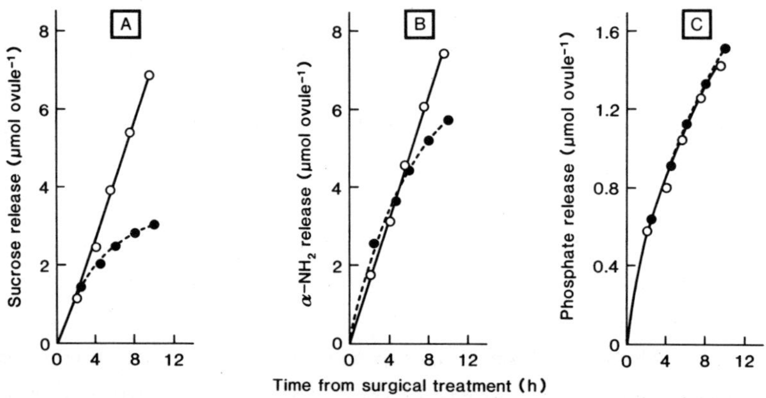

FIG. 23. Secretion of sucrose (A), amino compounds (B) and Pi (C) from attached (open symbols) or detached seedcoats of developing pea seeds (cv. Marzia). The seeds were approaching maturity. From Wolswinkel & Ammerlaan (1985a).

* * * * * *

6.3.2 Secretion of Amino Acids and Amides

In Chapter 4, evidence was presented that asparagine is the nitrogenous solute in highest individual concentration in the phloem sap entering developing pea pods (Table 8). Yet only trace amounts of asparagine or aspartate can be found in embryo sac liquid from young pea seeds, where the most abundant forms of nitrogen are glutamine, homoserine and alanine, followed by serine and valine (Murray, 1983b, 1985). Developing seeds do not contain more than a trace of asparagine:pyruvate aminotransferase activity (Ireland and Joy, 1981), but they do possess asparaginase, which develops first in the seedcoats, then in the embryo (see Table 3; Fig. 24). Both asparaginase and aspartate:oxoglutarate aminotransferase activities are maximal in young seedcoats, about 2 weeks after full blossom (Fig. 24). In combination with a restricted rate of unloading of asparagine from phloem in seedcoats, these two enzyme activities can account for the selective removal of asparagine from the range of amino compounds secreted into the embryo sac. Seedcoats from developing pea seeds exhibit up to 10-fold greater aminotransferase activity (fresh weight basis) than pea leaves, with the isoenzymes from mitochondria and the cytosol predominant.

Earlier it was assumed that the oxaloacetate derived from asparagine would probably be utilized primarily for the synthesis of malate and other organic acids secreted into the embryo sac (Murray and Kennedy, 1980). However, pulse-chase studies with [^{14}C]asparagine and other labelled amino compounds indicated that the major metabolic fates of asparagine in young pea seedcoats are incorporation into newly synthesized protein, and conversion to glutamine, alanine, valine and sucrose; malate becomes more significant as a product after the embryo sac liquid has been resorbed (Fig. 25). Within the hull, label acquired from pulse-fed [^{14}C]asparagine remains associated mainly

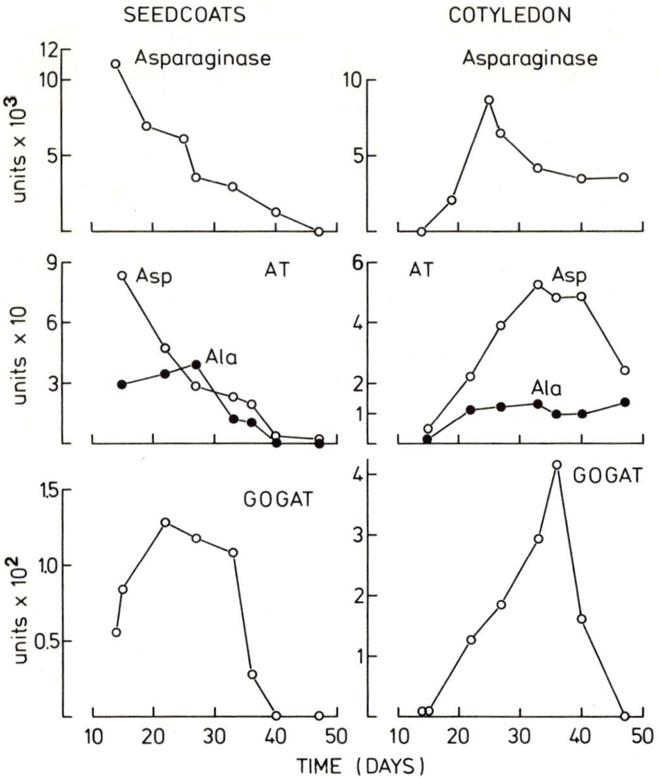

FIG. 24. Changes in the activities of asparaginase, aspartate:oxoglutarate aminotransferase, alanine oxoglutarate aminotransferase (●) and GOGAT, in seedcoats and a single cotyledon from developing pea seeds (cv. Melbourne Market) - note the scale differences along each ordinate. From Murray and Kennedy (1980).

* * * * * *

with asparagine (Table 12).

To confirm that the metabolic conversion of asparagine to secreted metabolites takes place in the seedcoats, and not in some more distant organ such as the subtending leaf an experiment was performed with this leaf cut off (see Section 4.4.4). The removal of the subtending leaf made

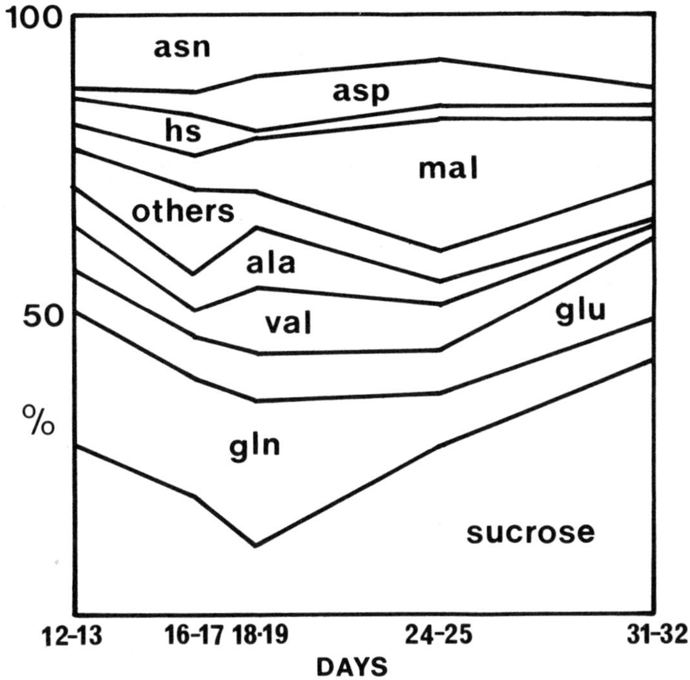

FIG. 25. Distribution of label in ethanol-soluble metabolites from the seedcoats 24 h after pulse-feeding [^{14}C]-asparagine via the cut stem of fruiting pea plants. hs = homoserine; mal = malate. From Murray & Cordova-Edwards.

* * * * * *

no substantial difference to the distribution of label in the three major secreted products of asparagine metabolism - glutamine, alanine and sucrose (Table 18). The data are therefore entirely consistent with the conclusion that most asparagine metabolism in the young pea pod takes place selectively within the seedcoats.

The conversion of asparagine to glutamine involves co-operation of the mitochondria with the cytosol, or other plastids (Fig. 26). This scheme envisages that asparaginase acts in the cytosol (Ireland and Joy, 1983) and that aspartate is then exchanged by the mitochondria for glut-

Table 18.

Distribution of Label in Metabolites in the Embryo Sac Liquid of Seeds from a Pea Pod from a Plant Fed [^{14}C]asparagine via the Cut Stem, With or Without the Subtending Leaf Attached (radioactivity as % of total).

	With leaf[a]		Without[b]
Compound	12-13 d	16-17 d	15-16 d
Alanine	15.10	16.93	16.52
Aspartate	0.43	1.14	0.72
Asparagine	5.60	3.64	1.46
Sucrose	7.27	18.28	14.75
Glutamate	1.22	1.07	1.02
Glutamine	52.42	43.37	43.00
Homoserine	5.32	5.80	1.58
Threonine	2.06	2.95	3.03
Serine	1.31	1.17	1.24
Glycine	ND	ND	0.51
Valine	8.79	4.50	3.08
Others	0.48	1.15	13.09

[a] 1981 season; homoserine omitted from chase solution

[b] 1985 season. From Murray & Cordova-Edwards (1984b) and Murray (1987).

* * * * * *

amate formed by the pathway indicated (Fig. 26). Export of glutamate from the mitochondria could also be a critical factor in regulating the syntheses of alanine and valine (Figs 25, 27). Studies of pea leaf mitochondria by Bryce and Wiskich (1985) indicate that pH, oxaloacetate supply and glutamate concentration could all influence malic enzyme in the direction of pyruvate synthesis (step 6 in Fig. 27). If glutamate were to accumulate in the mitochondria instead of being exported, then a reversal of the malate dehydrogenase reaction would be favoured, and C_4-units would be channelled through the Krebs cycle instead.

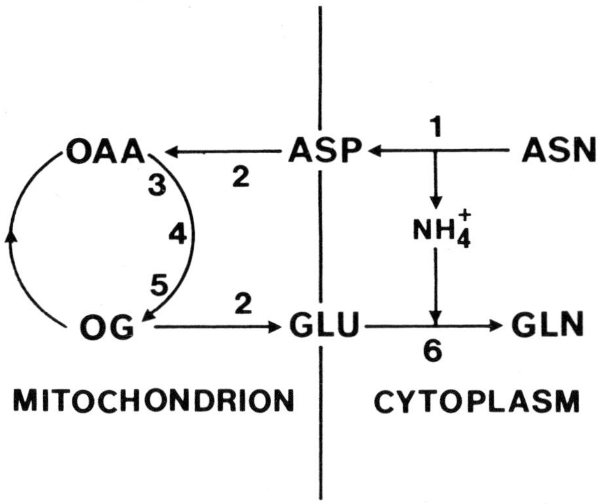

FIG. 26. The conversion of asparagine to glutamine in young pea seedcoats proceeds via aspartate, oxaloacetate (OAA), oxoglutarate (OG) and glutamate. Enzyme key: 1, asparaginase; 2, aspartate:oxoglutarate aminotransferase; 3, condensing enzyme; 4, aconitase; 5, isocitrate dehydrogenase; 6, glutamine synthetase. From Murray & Cordova-Edwards, 1984b.

* * * * * *

Events that might lead to this diversion include a declining asparaginase activity (Fig. 24) and hence aspartate availability, or a change in the abundance or specificity of dicarboxylate transporters in the mitochondrial membranes. Mitochondrial function in seedcoats is clearly important throughout development. On a dry weight basis, the oxygen consumption of seedcoats attributable to respiration is at most stages considerably greater than that of the embryo, or of the whole seed (Bils and Howell, 1963). These authors isolated mitochondria from seedcoats by conventional methods, but it is now possible to study the properties of intact mitochondria isolated from protoplasts released from seedcoat parenchyma. Compartmentation

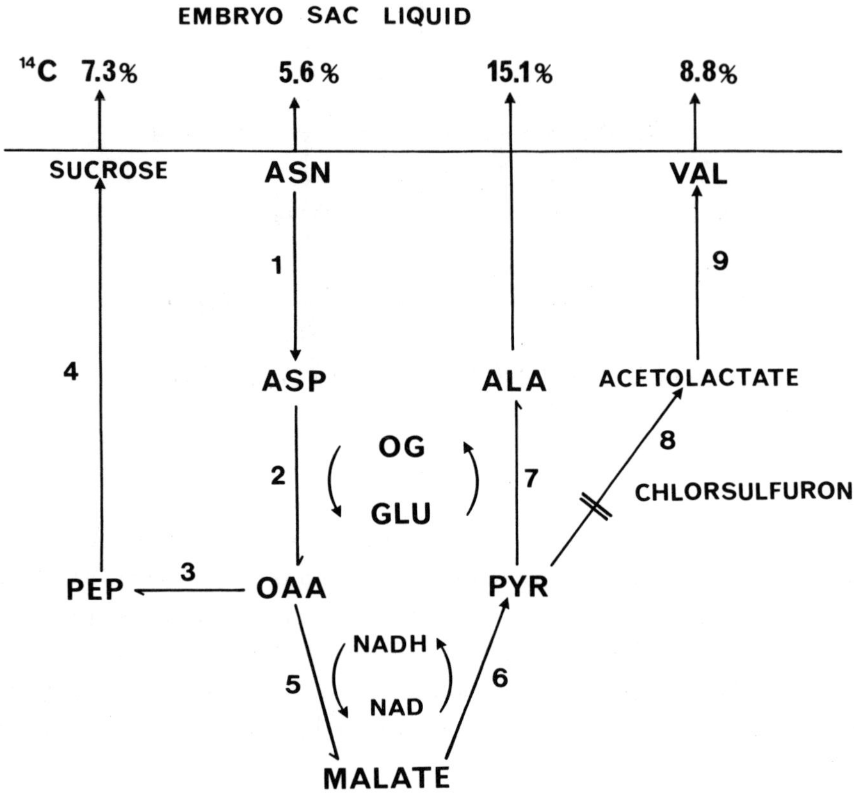

FIG. 27. Metabolic scheme to account for the conversion of asparagine to alanine and valine in young pea seedcoats together with the distribution of label in major metabolites of the embryo sac liquid (other than glutamine) for a 12-13-day old pod following [^{14}C]asparagine feeding via the cut stem (see also Table 18). Enzyme key: 1, asparaginase; 2, aspartate:oxoglutarate aminotransferase; 3, PEP carboxykinase; 4, gluconeogenesis and sucrose synthesis; 5, malate dehydrogenase; 6. malic enzyme; 7, alanine:oxoglutarate aminotransferase; 8, acetolactate synthase; 9, enzymes of valine synthesis. From Murray (1987).

* * * * * *

of the metabolic pathways outlined in Figures 26 and 27 might then be more precisely defined, but in their essential features, these new pathways have already been confirmed.

Sulfonylurea herbicides specifically inhibit acetolactate synthase from pea seedlings (Ray, 1984), and if such compounds could gain access to seedcoats, might be expected to inhibit the synthesis of the branched-chain amino acids valine, leucine and isoleucine in the seed *in vivo*. The incorporation of label into valine secreted into the embryo sac was inhibited by 63% when 1 μM chlorsulfuron was included in the chase solution following pulse-feeding of [^{14}C]asparagine to a fruiting pea plant via the cut stem (Murray, 1987). This treatment did not affect the total label among solutes in embryo sac liquid, only the distribution of label.

When [^{14}C]alanine was supplied to a fruiting pea plant via the cut stem, only 4% of the label in metabolites secreted into the embryo sac was still associated with alanine (Murray, 1986a). The transmitted label was mainly in glutamine (27% of the total) and sucrose (60%). In a similar experiment with the subtending leaf excised, the proportion of label represented by sucrose fell by about half - the proportions in glutamine and sucrose were almost equal (Section 4.4.4). The conversion of alanine to sucrose must therefore be taking place partly in the subtending leaf, and partly in the stem, prior to phloem loading. The hull is unlikely to contribute to the conversion of alanine to sucrose, because when a developing pod was fed via the peduncle, most of the label in solutes entering the embryo sac was in glutamine (72%). Hence the pod wall is the main site for conversion of alanine to glutamine.

It follows that almost all of the alanine secreted into the embryo sac must be formed *de novo* in the seedcoat parenchyma, mainly from asparagine (Fig. 27). The abundance of alanine in the seedcoat secretion from young broad bean seeds (Wolswinkel and de Ruiter, 1985) and in embryo sac liquid from cowpea seeds (Peoples et al., 1985a) suggests that this metabolic pathway is widespread.

Threonine was observed to increase in relative abundance in embryo sac liquid from pea seeds at the same time as the contents of homoserine decreased (Murray, 1983b). In pulse-chase experiments with [^{14}C]homoserine (Section 5.3.1), this suspected conversion of homoserine to threonine was confirmed. De Ruiter et al. (1984) analysed the seedcoat secretion at a 3-week stage in a different cultivar of pea. Threonine ranked third in concentration after alanine and glutamine, but homoserine had diminished to represent only a minor proportion of all the amino solutes secreted (Table 19). Once again a shift in seedcoat metabolism has led to a major change in the composition of the spectrum of potential nitrogen sources made available by the seedcoats for uptake by the developing embryo. A significant difference between pea cultivars is evident in the prominence of arginine in the secretion from seedcoats of cv. Marzia (Table 19), but not from cv. Melbourne Market (Fig. 28).

Developing cowpea seeds provide an interesting comparison with peas. Of the total nitrogen in solutes released to the embryo sac liquid, histidine accounts for 30-55%; alanine, proline and glutamine are the next most abundant forms, whereas asparagine accounts for only 5-10% (Peoples et al., 1985a). It is noteworthy that NH_4^+ and the ureides, allantoin and allantoic acid, are very minor constituents of the embryo sac liquid, and of the seedcoat secretion monitored at an intermediate stage of development. In soybean also the ureides are not transmitted to the embryo sac to any significant extent (Hsu et al., 1984; Rainbird et al., 1984). Coker and Schaefer (1985) have confirmed that cut pieces of seedcoats from developing soybean seeds

Table 19.

Amino Compounds in the Seedcoat Secretion from Young Pea Seeds (cv. Marzia; 3 weeks post-anthesis).

Compound	mol %	mol N %
Alanine	23.6	16.4
Glutamine	22.3	31.0
Threonine	10.0	7.0
Serine	7.0	4.9
Arginine	5.4	14.9
Valine	5.3	3.7
Glutamate	5.2	3.6
Leucine	3.2	2.2
Isoleucine	3.0	2.1
Homoserine	2.2	1.5
Lysine	2.2	3.0
γ-Aminobutyrate	2.2	1.5
Phenylalanine	1.7	1.2
Methionine	1.5	1.0
Asparagine	1.4	1.9
Glycine	1.3	0.9
Tyrosine	1.0	0.7
Histidine	0.8	1.6
Aspartate	0.8	0.6
Ornithine	0.2	0.3

The collection period was 230 min. Data of de Ruiter, Schuurmans and Kollöffel (1984).

* * * * * *

are better able to metabolize labelled allantoin than are cotyledons. *In vivo*, it is not certain whether substantial amounts of ureides are imported by seedcoats; glutamine, asparagine and amino acids are more abundant than ureides in pedicel phloem sap from cowpea (Peoples et al., 1985b).

Seedcoat secretions collected about midway through development from cowpea seeds indicate that glutamine and

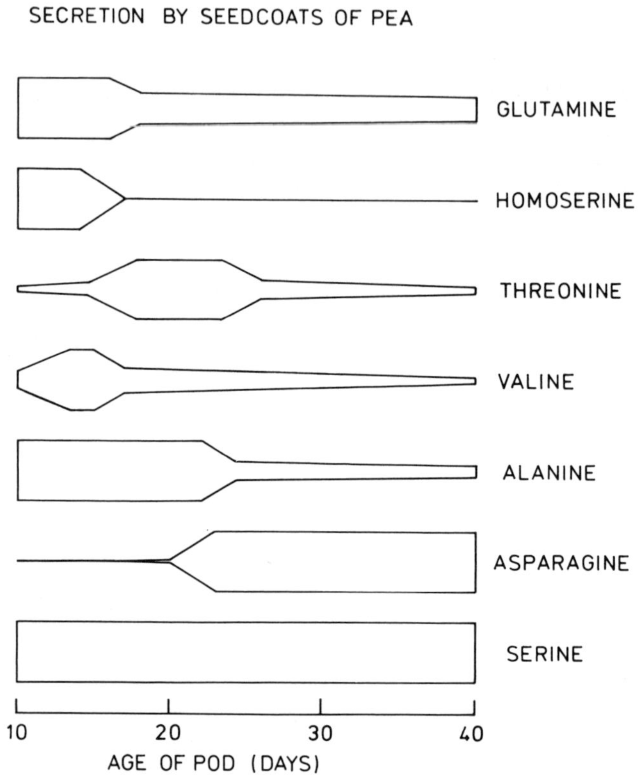

FIG. 28. Schematic diagram of the changes in relative abundance of the major nitrogenous solutes secreted by the seedcoats of developing pea seeds (cv. Melbourne Market).

* * * * * *

asparagine have become the principal secreted forms of nitrogen, followed by histidine (declining from its early predominance) and arginine, then serine and alanine (Peoples et al., 1985a). Glutamine and asparagine are also the most abundant nitrogenous solutes secreted by soybean seedcoats at the stages examined by Rainbird et al. (1984) and by Gifford and Thorne (1986). Amino compounds (mainly amides) contribute 30-50% as much mass as sugars to the seedcoat secretions of soybean. A molar ratio of unity for amino compounds:sucrose is typical also of seedcoat secretions from developing pea seeds (Fig. 23). This ratio is about an order of magnitude greater than that prevailing in phloem sap from legume pods or fruit stalks (Chapter 4).

The diminishment of asparagine content in favour of alternative nitrogenous solutes in seedcoat secretions is often found in young legume seeds. And yet asparagine, with glutamine, is a major nitrogen source for the embryo later in development. Why is the metabolism of young seedcoats arranged in such a way as to withdraw asparagine?

Two possibilities suggest themselves. One advantage may be the liberation of divalent mineral cations associated with asparagine in natural chelates (Chapter 4). The ability of asparagine to form complexes involving two molecules per cation has been demonstrated in studies of pea pods treated with exogenous Cu^{2+} (Cruickshank et al., 1987). The role of such chelates in translocation has already been discussed (Section 4.3). There is no doubt that mineral cations are released from the forms in which they are obtained by fruit tissues and seedcoats, as adequate contents of all the essential mineral cations are present in seedcoat secretions (Section 6.1.3).

Another possible rationale for asparagine metabolism in the seedcoats concerns the operation of the Davies pH stat referred to by Kennedy (1986; p. 92). The PEP carboxylase reaction generates protons, required in seedcoat parenchyma for the reoxidation of NADPH or NADH (e.g. the conversion of oxaloacetate to malate, step 5 in Fig. 27), and for secretion to the embryo sac *per se*. The PEP carboxylase reaction simultaneously replaces a C_3-unit with a C_4-unit, thus serving an anaplerotic function for a modified Krebs cycle committed to the withdrawal of C_5-units to provide glutamate, as shown in Figure 26. The effect of PEP carboxylase activity on carbon balance *in vivo* may not be pronounced. There are numerous opportunities also for decarboxylation (e.g. steps 3 and 6, Fig. 27, and normal operation of the Krebs cycle). That concentrations of CO_2 attain such high values in the gas space of the pod (e.g. up to 4.3%, Harvey et al., 1976) is evidence that both the intact seeds and the inner epidermis of the hull lack the facility to recover respiratory losses of CO_2 efficiently.

PEP carboxylase activity appears necessary to support all aspects of asparagine metabolism in seedcoats, including its conversion to glutamine (Fig. 26), alanine, valine and malate (Figs 25, 27). Thus the metabolism of asparagine during the early development of the seedcoats may afford a complex mechanism for generating sufficient protons. This cannot be the sole mechanism for balancing proton requirements of metabolism and secretion, as evidenced by an increasing transmission of asparagine to the embryo sac later in development. Surplus protons are required throughout the course of seed development (Section 7.5).

An incidental comment is called for on the transmission of sulphur by seedcoats. Although glutathione and homoglutathione are the assimilated forms of sulphur arriving in fruit tissues and seedcoats via the phloem (Section 4.4.2), the only sulphur-containing metabolite identified so far in the seedcoat secretion is methionine (Table 19). The concentration of this form of sulphur is so low that it does not appear adequate to meet the needs of storage protein synthesis in developing embryos. This is discussed further in Section 8.2.

6.3.3 Secretion of Sugars

Studies of evacuated seedcoats show that sucrose is the predominant sugar secreted into the embryo sac. Nevertheless, profiles of the composition of embryo sac liquid often indicate substantial contents of reducing sugars, particularly glucose (Section 6.1.3). These observations raise several questions. Are these reducing sugars derived from sucrose following secretion by the seedcoats, possibly in the endosperm cytoplasm, or in the central vacuole of the embryo sac itself? Or is there specific secretion of hexoses by seedcoats in addition to the predominant release of sucrose?

Evidence supporting the latter hypothesis has been gained by the analysis of both seedcoats and secretions. Gifford and Thorne (1985) determined the glucose and sucrose contents separately for cotyledons and seedcoats of developing soybean seeds (Table 20). The molar ratio of sucrose to glucose was 6.6-6.9, and this did not vary according to the time of day the sample was collected. This study, together with that of Fader and Koller (1985), establishes that a pool of free glucose is normally present in soybean seedcoats, and so might be available for secretion. That secretion of glucose occurs was confirmed by Gifford and Thorne subsequently (1986; see Figs 30, 31).

Table 20.

Apparent Concentrations of Sucrose and Glucose in Seedcoat and Cotyledon Tissue Water (mM). Developing Soybean Seeds were Extracted at Different Times of the Day.

Time	Seedcoats:		Cotyledons:	
of day	Sucrose	Glucose	Sucrose	Glucose
11 a.m. - Noon	103.5	15.0	83.0	8.3
6 p.m.	89.6	13.6	90.4	12.0
7 a.m.	97.3	14.8	85.5	10.5

Data of Gifford and Thorne (1985).

* * * * * *

Wolswinkel and Ammerlaan (1985a) measured separate rates of secretion for glucose and sucrose from seedcoats of developing broad bean seeds. The rate of sucrose release is always greater than the rate of glucose release. Pea seedcoats also secrete both sugars (Wolswinkel and de Ruiter, 1985). The production and secretion of glucose by seedcoats is therefore a general occurrence.

The action of invertase cannot be invoked to explain the origin of any free glucose pool in pea seedcoats or embryo sac liquid (L. Copeland, personal communication), and the activity of α-glucosidase in pea seedcoats is very weak (Murray, 1983a). The immediate precursors of glucose in seedcoats could therefore be glucose-6-phosphate and glucose-1-phosphate. The acid phosphatase from pea seedcoats acts against glucose-6-phosphate at 24% of the rate against the model substrate *p*-nitrophenylphosphate (Fig. 20), and could well play a part in generating both glucose and Pi *in vivo*. This role is consistent with the suggestions of Turner and Turner (1960).

All of the hexose and hexose phosphate pools of seedcoat parenchyma must be derived from a small proportion of the sucrose imported from phloem sieve elements and retained by the seedcoats. By undergoing the sucrose synthase reaction in reverse, as in developing fruit tissues (Section 5.4), sucrose would yield both UDP-glucose and fructose. The UDP-glucose derived in this way could generate glucose -1-phosphate by the subsequent action of UDP-glucose pyrophosphorylase. These reactions have been studied in young expanding leaves (Clausson et al., 1985a,b), developing broad bean cotyledons (Hawker, 1971), and developing pea embryos (Edwards and ap Rees, 1986a,b), but not yet in seedcoats.

Fructose, which is not secreted in significant quantities from legume seedcoats, might be converted rapidly to glucose phosphates, first via fructokinase (Copeland et al. 1978), then via phosphoglucoisomerase and phosphoglucomutase, both of which are active as two major isoenzymes in pea seedcoats (Murray and Ayre, 1987).

CHAPTER 7
Phloem Unloading in Seedcoats

Chapter 6 has documented the changes imposed on solutes as they pass through seedcoat parenchyma prior to their secretion into the embryo sac. Transmission, modification and secretion of metabolites are distinct from the process of phloem unloading. It is important to make this distinction, in order to trace the cellular pathways of solute transfer through seedcoat parenchyma, and to determine which of the factors regulating seedcoat function do so by exerting their influence on the unloading system.

7.1 Delivery and Allocation of Solutes via Phloem

It is generally accepted that developing seeds obtain their entire supply of nutrients from the phloem. However, we have very limited information on how seeds exercise a priority in gaining individual solutes in competition with the fruit walls, or other developing parts of the plant. Each individual solute carried to the fruit in phloem may be retained within the veins, distributed within the fruit or allocated to the seeds, in proportions that differ with solute and stage of development (Section 5.2.1).

Part of the explanation for this may be that the development of phloem sieve elements in terminal regions of the vasculature involves the synthesis of specific protein transport complexes in the plasmalemma. Carriers responsible for mediating the transfer of sucrose are probably distinct from those responsible for transporting amino compounds, and their numerical frequencies within the plasmalemma might also vary with stage of development. These suppositions are based on studies of phloem loading and unloading systems in vegetative parts of the plant, e.g. in leaves (Servaites et al., 1979; Schmalstig and Geiger, 1985, 1987).

7.2 Pathways of Nutrient Transfer from Phloem

Thorne (1985) has recently described the pathways for solute transfer following unloading from the phloem in developing grains of tropical and temperate cereals. Symplastic transfer is a common feature of these pathways, up to certain regions where solutes are required to enter the apoplast, e.g. in the placenta-chalaza of *Zea mays*, where cell-wall invertases hydrolyse a high proportion of incoming sucrose at an early stage of development (Felker and Shannon, 1980; Porter et al., 1985).

In temperate cereals such as wheat, unloading occurs along the full length of a single vascular bundle running through the pericarp. Transfer takes place through the symplast up to and including the cells of the nucellus (Oparka and Gates, 1981a,b; 1982), which serves as a secretory tissue. Sucrose is not hydrolysed immediately after unloading (Jenner, 1974; Sakri and Shannon, 1975); following secretion, it accumulates at an approximate concentration of 230 mM in the endosperm cavity (Gifford and Bremner 1981a). Sucrose and glutamine are delivered, unloaded and transmitted by broadly similar mechanisms (Donovan et al., 1983).

A pulse of [^{14}C]sucrose arriving in a developing legume seed accumulates in the seedcoats prior to any onward transmission to the embryo sac (Thorne, 1980; Patrick and McDonald, 1980). This label becomes uniformly distributed throughout the innermost layers of parenchyma, as demonstrated by autoradiography (Thorne, 1980). Furthermore, most of the label secreted to the embryo sac remains associated with sucrose (Patrick, 1983).

Most seedcoat cells are not in immediate contact with phloem sieve elements, and must receive imported solutes via symplastic connections (plasmodesmata) or by uptake from the apoplast. The parsimonious way to explain the distribution of label described above is by considering that sucrose is imported by cells adjacent to phloem sieve elements, then transferred from cell to cell by plasmodesmata (Gunning and Robards, 1976). Phloem sieve elements are normally connected to adjacent companion or parenchyma cells by plasmodesmata (Evert, 1984; Esau and Thorsch, 1985). This holds true for seedcoat sieve elements (Thorne, 1981, 1985; Offler and Patrick, 1984, 1986), with plasmodesmata connecting even the most remote seedcoat cells back to cells adjacent to phloem.

Ultrastructural observations indicate that the transmission of sucrose through the parenchyma of seedcoats of developing bean seeds could be mainly or entirely via the symplast (Fig. 29). Pitfields (clusters of plasmodesmata) are most abundant in parenchyma cell walls aligned at right-angles to the probable main directions of solute transfer, both lateral (circumferential) and radial (Table 21). Estimates of sucrose fluxes through cross-sectional areas of plasmodesmata confirm that a symplast route would be sufficient for the transmission of solutes through the seedcoats at observed or inferred rates (Table 21). Similar observations and calculations have been made for the seedcoats of developing broad bean seeds (Offler and Patrick, 1986).

Yeung (1983) suggested that in bean seedcoats, the inner 'branched' (or 'branch' parenchyma, Fig. 29) are the cells responsible for secretion. This assessment agrees with that of Offler and Patrick (1984; see Fig. 29). In some

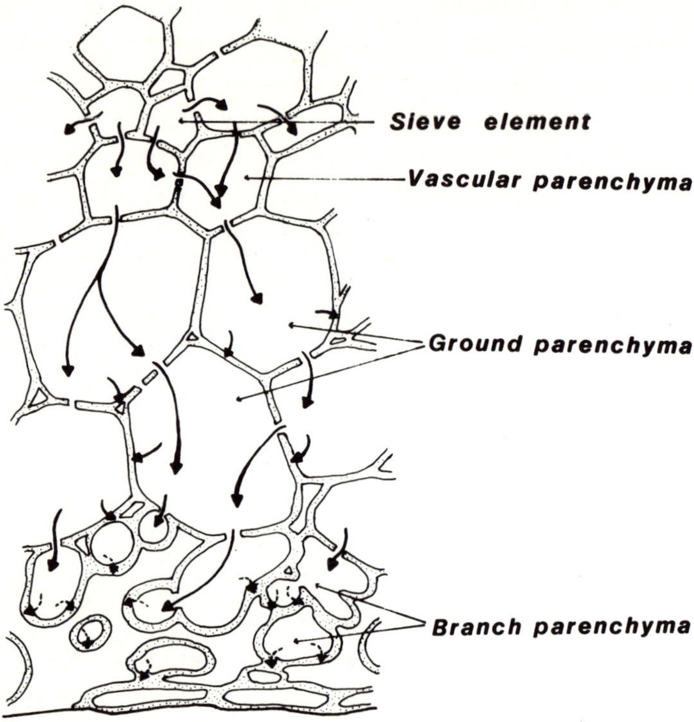

FIG. 29. Diagram of the cell types comprising the short-distance pathway for sucrose transfer through the seedcoat of developing seeds of *Phaseolus vulgaris*. Solid arrows, predominant cellular pathway for sucrose transfer; dashed arrows, transfer through plasmalemma. From Offler and Patrick (1986).

* * * * * *

other species, such as broad bean (Offler and Patrick, 19-86), cells of the innermost layer flanking the embryo sac display the wall ingrowths typical of transfer cells, an adaptation believed to enhance the surface area of membranes housing solute transporters.

The retention of sucrose by the protoplasts of seedcoat parenchyma must be regulated by compartmentation and meta-

Table 21.
Plasmodesmatal Frequencies, Surface Areas of Shared Wall, and Estimated Sucrose Flux per Unit Cross-Sectional Area of Plasmodesma for Contiguous Cell Types in the Radial Pathway of Sucrose Transfer Through the Seedcoats of Bean *(Phaseolus vulgaris)*.

Types of contiguous cells	Plasmo-desmatal frequency (wall)[a]	Shared wall surface area[b]	Plasmo-desmatal frequency (coats)[c]	Sucrose flux[d]
Sieve element[e]/ vascular parench.	0.41	0.044	0.018	8.40×10^6
Vascular parench./ vascular parench.	2.95	0.18	0.53	2.23×10^5
Vascular parench./ ground parench.	1.72	0.06	0.10	1.35×10^6
Ground parench./ ground parench.	1.02	7.66	7.81	1.62×10^4
Ground parench./ branch parench.	1.37	0.17	0.23	4.26×10^5
Branch parenchyma/ branch parenchyma	1.49	0.73	1.09	1.15×10^5

[a] $\times 10^{-8}$ cm^{-2} of wall [b] cm^2 cm^{-2} of seedcoat

[c] $\times 10^{-8}$ cm^{-2} of seedcoat

[d] pmol sucrose s^{-1} cm^{-2} plasmodesmatal area

[e] includes companion cells. From Offler & Patrick (1984).

* * * * * *

bolism if most of the sucrose passing through the protoplasts of these cells ultimately reaches the cells responsible for secretion. How can this be accomplished by internal compartmentation, since the protoplasts are all interconnected? Such difficulties with the symplast proposal have led P. Wolswinkel and his colleagues to propose an alternative explanation of solute transfer through seedcoats (Wolswinkel, 1985a,b; 1987; Wolswinkel and Ammerlaan

1985a,b; Wolswinkel et al., 1983, 1986). This involves release of sucrose and amino compounds from the terminating phloem sieve elements to the apoplast, combined with recapture of solutes by seedcoat parenchyma cells with "cotyledon characteristics" (Wolswinkel, 1985a).

A major difficulty with this proposal is the irreversible nature of seedcoat secretion. Once committed to the apoplast communicating with the embryo sac, major solutes like sucrose do not again penetrate the seedcoat parenchyma - at least not in sufficient quantity to support the 'recapture' hypothesis. When [^{14}C]sucrose is placed in evacuated attached soybean seedcoats, not more than 0.5% of the total radioactivity is taken back through the seedcoats into other plant parts (Bennett et al., 1984). Using attached pea seedcoats and the short-lived isotope ^{11}C as a label, Minchin and McNaughton (1986) have confirmed that "little or no sucrose is absorbed into the seedcoat tissue from the seedcoat apoplast". The technique devised by Minchin and McNaughton involves continuous monitoring of the solution circulating in the seedcoat cup, and is extremely sensitive.

The data on solute uptake by seedcoats which have been taken to support the apoplast recapture hypothesis have been obtained with excised seedcoat halves incubated in a liquid medium *in vitro*. It is possible that solute uptake has occurred through cut surfaces that might not provide entry for the same solutes *in vivo*. Since the parenchyma of seedcoats are capable of precise metabolic transformation of nitrogenous solutes delivered from phloem prior to secretion of a modified spectrum of amino compounds (Section 6.3.2), and the same cells are capable of generating a pool of glucose specifically for secretion (Section 6.3.3), the evidence favouring a symplastic mode of transport for solutes through seedcoat parenchyma is overwhelming. It may be conceded that Wolswinkel's seedcoat cells with 'cotyledon characteristics' might be confined to the cell layer in immediate contact with phloem sieve elements - but it is difficult to see how this hypothesis could be tested unequivocally.

7.3 Experimental Manipulation of Phloem Unloading

When metabolic inhibitors or other solutes (e.g. K^+) are applied in solution to seedcoat cups and the efflux of sucrose or any other metabolite is inhibited or stimulated the immediate conclusion that phloem unloading itself has been inhibited or stimulated by the treatment is not justified. There are several possible sites of effect that might ultimately feed back to an inhibition or stimulation of phloem unloading. It has been impossible in many published reports to discriminate between direct and indirect effects of experimental treatments. A given compound might affect phloem unloading, symplastic transfer, or secretion through the plasmalemma; according to concentration, it may act directly by affecting membrane components, or indirectly, by inhibiting the mitochondrial production of ATP. Nevertheless, several authors believe that specific effects on phloem unloading in seedcoats have now been identified unequivocally.

7.3.1 Effects of Anoxia

The oxygen requirements of developing seed tissues that lack chloroplasts are probably generated substantially by chloroplasts in nearby supply organs, and the oxygen so produced conducted to the seedcoats, if necessary, via the phloem. Gifford and Bremner (1981b) concluded that the photosynthetic production of oxygen in the inner green pericarp layer of wheat grains has a "diffusional advantage", and given the impermeable fruit-coats of the grain, is the most important source of oxygen supporting sucrose transmission, and starch synthesis in the endosperm.

In legumes, developing seeds quickly develop a waxy cuticle (Hardham, 1976; Harris, 1983), which is impermeable to gas uptake from the closed interior of the pod. This is confirmed by the failure of developing pea seeds to fix $^{14}CO_2$ from the pod atmosphere (Section 5.3.1). Etiolated pea pods lacking functional chloroplasts in the hull and seedcoats grow more slowly than diurnally-illuminated pods (Section 5.3). One reason for this difference could be a lack of oxygen, necessary for all aspects of carbon and nitrogen metabolism involving mitochondria (e.g. Figs 26, 27), and for maintaining adequate supplies of ATP for solute transport.

Thorne (1982a) examined the interaction of temperature and O_2 availability on the accumulation of ^{14}C-labelled solutes by developing soybean seeds. He showed that when developing seeds in their pods are completely deprived of exogenous O_2, ^{14}C-labelled photosynthate (mainly sucrose) does not proceed beyond the confines of the seedcoat vascular bundles. Unlabelled embryos isolated from such seeds and incubated under O_2-free nitrogen are still capable of sucrose uptake under these conditions, and so would have taken up [^{14}C]sucrose *in vivo* if any had been unloaded, transmitted and secreted under anoxic conditions. Since the labelled solutes were retained inside the vascular bundles of the seedcoats and could not be subject to further transport, metabolism and secretion, an oxygen requirement associated specifically with the energy requirements of phloem unloading can be recognized.

7.3.2 Equilibration and Estimation of Import

Gifford and Thorne (1985, 1986) developed a way of estimating the rate of phloem unloading within seedcoats, based on observations that the time-course of solute efflux from attached soybean seedcoats exhibits three phases (Figs 30-32). The first phase occupies only a few minutes, and represents rapid apoplast washout or purging (see also Patrick, 1984; Minchin and McNaughton, 1986; papers by Wolswinkel). The secretion rates for sucrose, glucose and total amino compounds are all most rapid initially (Figs 30-32; see also Wolswinkel and Ammerlaan, 1985a).

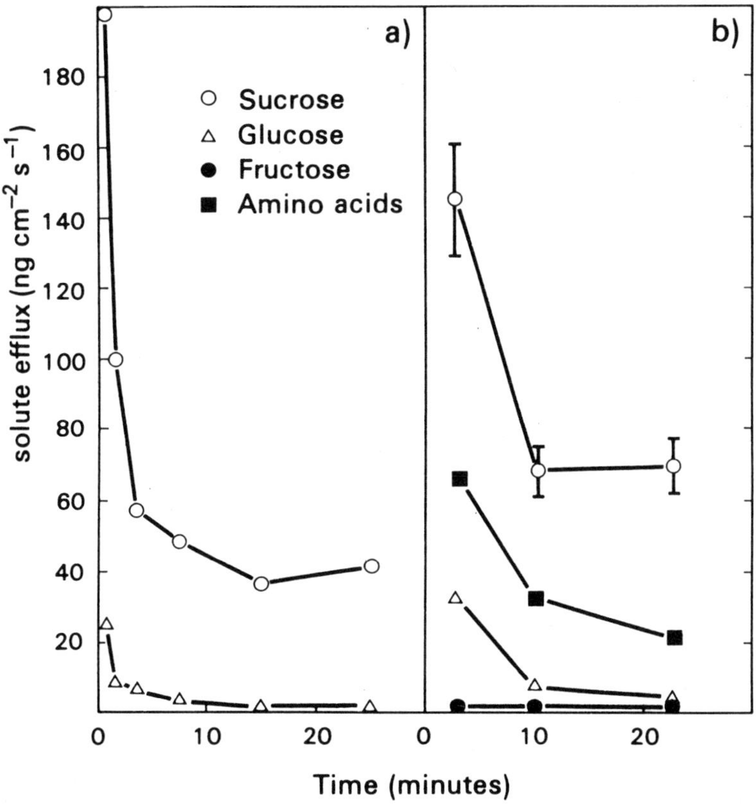

FIG. 30. Efflux of sucrose, glucose, fructose and total amino compounds for 25 min after placing oxygenated deionized water into attached soybean seedcoat cups. Values are expressed per unit area of seedcoat surface. From Gifford and Thorne (1986).

* * * * * *

In the second phase, a minimum efflux of solutes occurs about 2 to 3 hours from the beginning of the experiments

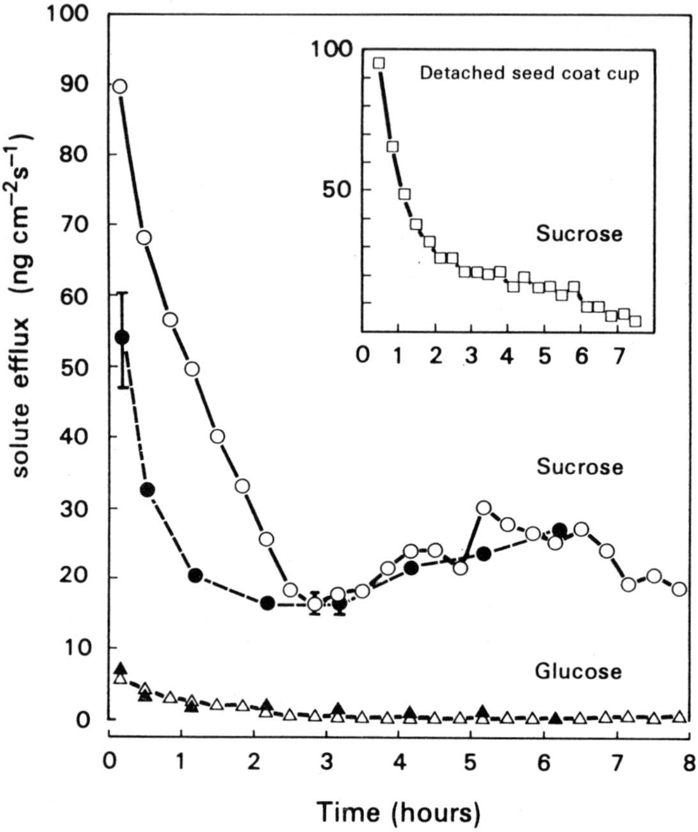

FIG. 31. Long-term efflux of sucrose and glucose from attached soybean seedcoats (two experiments distinguished by open vs. black symbols) and of sucrose from a detached seedcoat (inset). From Gifford and Thorne (1986).

* * * * * *

(Figs 31, 32). In a typical third phase, the rate of release increases again, eventually declining. Gifford and Thorne (1986) interpret the second phase as a period of equilibration between the sucrose pool within the symplast of the seedcoat parenchyma and the trapping solution in the embryo sac. This is maintained at very low solute concentrations by periodic replacement (every 20 minutes).

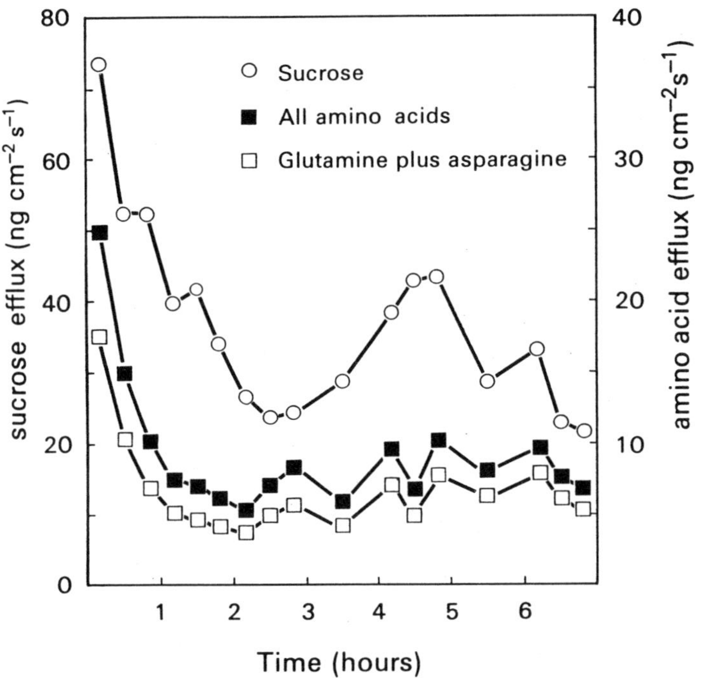

FIG. 32. Long-term efflux of sucrose and amino compounds from a single attached soybean seedcoat. From Gifford and Thorne (1986).

* * * * * *

The third phase, following this equilibration, should be characterized by rates of secretion that closely match the rates of sucrose import via the funiculus, then equivalent to the rates of unloading within the seedcoats. The amount of sucrose imported through the funiculus, S_u, is given as

$$S_u = S_e - (S_{ci} - S_{cf}) + r_c$$

where S_e is the total amount of sucrose accumulated in the liquid trap, S_{ci} and S_{cf} are the initial and final sucrose contents of seedcoat tissues respectively, and r_c is the amount of sucrose equivalent to hexose consumed in respiration, which can be measured as O_2 uptake with an electrode. It is assumed for the purpose of calculation that there is no significant resynthesis of sucrose from hexose phosphates during the period of the experiments.

Estimates of sucrose import (= sucrose unloaded from phloem) and sucrose efflux per seedcoat agree fairly well when either water or buffer at pH 6.3 is used as the trap solution (Table 22). Osmotic solutes and Ca^{2+} have no pronounced effects at the concentrations tested. However, some direct effects on sucrose import can be distinguished for certain compounds, especially those affecting supplies of ATP. Potassium cyanide (100 μM) has no significant effect on 'steady state' sucrose efflux, but lowers the import of sucrose by 60-70% (Table 22). This concentration of CN^- inhibits seedcoat respiration by only 45%; presumably cyanide-insensitive respiration is capable of supporting the lower rate of sucrose import observed. In contrast to CN^-, the uncoupler trifluoromethoxy(carbamylcyamide) phenyl hydrazone (FCCP) blocks sucrose import totally, a result consistent with the elimination of mitochondrial ATP formation. Fusicoccin, an ATPase stimulant, enhances both secretion and import of sucrose, a result that could be explained if the principal effect of fusicoccin were on sucrose/H^+ cotransport mediated by the plasmalemma of secretory parenchyma (Fig. 33).

The thiol reagent *p*-chloromercuribenzene sulfonate is generally non-permeating (Giaquinta, 1976; Rijven and Gifford, 1983), and is believed to inactivate plasmalemma carrier proteins by binding to their functional thiol group(s). The assumption that *p*CMBS is non-permeating is supported by its lack of effect on seedcoat respiration rate (Gifford and Thorne, 1986), although an earlier study noted approximately 10% inhibition of respiration in bean seedcoats (Patrick, 1983). The treatments with *p*CMBS that reduce sucrose import and efflux in soybean (Table 22) increase the rate of depletion of tissue sucrose, and there-

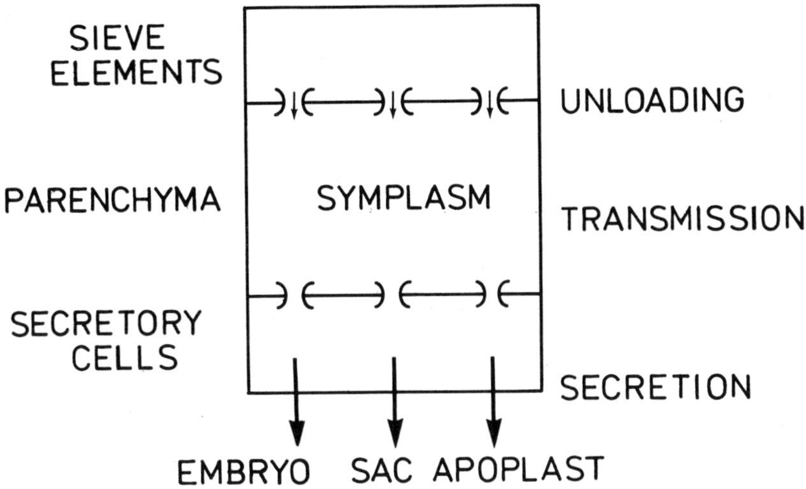

FIG. 33. Summary diagram of the processes by which solutes are imported from phloem then delivered through the inner seedcoat to the embryo sac.

* * * * * *

fore *p*CMBS may be acting principally at the plasmalemma of phloem sieve elements responsible for unloading sucrose into adjacent cells (Fig. 33). The reducing agent dithiothreitol stimulates or inhibits sucrose import according to concentration, but the oxidising agent H_2O_2 clearly inhibits the import of sucrose.

Both indoleacetic acid (IAA) and abscisic acid (ABA) elicit detectable increases in both sucrose efflux and import, reducing the loss of sucrose from within seedcoat parenchyma (Table 22). It cannot be claimed that these growth substances specifically affect phloem unloading. In attached pea seedcoats, the effects of ABA are variable and concentration dependent. Whether the effects obtained with exogenous IAA and ABA indicate that these or any related endogenous growth substances might have similar effects *in vivo* is another question, one that presently can-

Table 22.

Effects of Various Substances Placed in Solution in the Embryo Sac of Attached Soybean Seedcoats on Long-Term (3-5 h) Steady State Efflux of Sucrose (S_e) and the Calculated Rate of Sucrose Import (S_u)[a]. Values in parentheses are ±SE.

Addition	Whether buffered	S_u (ng s^{-1})	S_e (ng s^{-1})	Decline in tissue sucrose
Water control	No	19.5 (1.6)	21.7 (1.3)	0.55 (0.02)
Buffer[b], pH 6.3	Yes	23.0 (2.1)	23.6 (2.3)	0.49 (0.07)
Solutes:				
Mannitol, 150 mM	No	16.8 (4.7)	15.5 (3.3)	0.54 (0.04)
PEG 400, 150 mM	No	22.7 (3.0)	20.2 (2.6)	0.44 (0.06)
$CaCl_2$, 0.5 mM	No	18.7 (3.0)	19.2 (4.1)	0.45 (0.05)
Reagents affecting ATP supply:				
KCN, 100 μM	Yes	8.4 (2.4)	21.7 (2.7)	0.63 (0.05)
FCCP, 5 μM	Yes	-0.3 (1.7)	2.2 (0.1)	0.88 (0)
FCCP 20 μM	Yes	-5.9 (2.5)	NA[c]	0.93 (0.01)
Fusicoccin 20 μM	Yes	32.0 (2.4)	35.4 (2.6)	0.41 (0.03)

Redox reagents:							
*p*CMBS, 5 mM[d]	Yes	3.1	(1.3)	6.3	(1.1)	0.68	(0.01)
*p*CMBS 0.5 mM	Yes	6.0	(4.3)	5.9	(2.2)	0.68	(0.09)
Dithiothreitol, 1 mM	Yes	31.1	(5.3)	31.0	(1.6)	0.34	(0.09)
Dithiothreitol, 5 mM	Yes	6.3	(1.5)	13.4	(0.2)	0.67	(0.03)
H_2O_2, 0.03%	No	12.6	(5.5)	22.9	(4.8)	0.52	(0.11)
Plant growth substances:							
ABA, 40 μM	No	25.3	(0.9)	23.9	(1.2)	0.30	(0.05)
IAA, 40 μM	No	27.8	(4.9)	26.9	(3.9)	0.38	(0.01)

[a] the rates given are as ng sucrose per second per seedcoat

[b] 1 mM piperazine-N,N'-bis(2-ethanesulfonic acid)

[c] NA = not applicable, because efflux continued to decline

[d] applied for 5 min every hour, since a single brief exposure was ineffective

Data from Gifford and Thorne (1986).

not be answered (Section 7.6).

7.4 Osmotic Sensitivity of Solute Transmission

To what extent can the rates of phloem unloading and solute secretion be influenced *in vivo* by the osmotic potential of the solution in the embryo sac and apoplast of the enclosed embryo? The variety of experimental results obtained with just four legume species has made it difficult to answer this question. In soybean seeds the sucrose concentration in the apoplastic compartment is maintained close to 160 mM (Gifford and Thorne, 1985). In long-term experiments with attached seedcoat cups, the addition of higher concentrations of mannitol or polyethylene glycol to the bathing solution had no significant effects on sucrose import, sucrose efflux, or on the sucrose content of the seedcoat cells (Table 22). All of these parameters appeared to be unresponsive to changes in external solute concentration up to 500 mM mannitol (Thorne and Rainbird, 1983).

In short-term experiments with detached seedcoat halves from *Phaseolus vulgaris* seeds, Patrick (1984) found that the efflux of ^{14}C-labelled solutes (mainly sucrose) is greater if 100 mM sucrose is omitted from the solution placed in the seedcoat halves. This concentration of sucrose is similar to that found normally in the embryo sac and embryo apoplast in this species (Patrick, 1981).

The behaviour of broad bean and pea seedcoats following the removal of embryos is apparently different again. Wolswinkel and Ammerlaan (1984, 1986) and Minchin and McNaughton (1986) reported that increasing the external (embryo sac) concentration of solutes increases the sucrose efflux - up to concentrations of 1 M. Conversely, lowering the sucrose concentration (down to an apparent threshold of 300 mM) diminishes the rate of secretion of labelled sucrose (Minchin and McNaughton, 1986). These responses are not sucrose-specific; sorbitol and mannitol may substitute as external solutes.

It is possible to relate these various patterns of response to stage of development, rather than to species or laboratory. In *Vicia faba*, external solute concentrations necessary to support normal rates of secretion of sucrose and amino compounds from the seedcoats have been shown to differ according to stage of development. A concentration of 550 mM is necessary in early stages; 400 mM is inadequate for the early stages, but supports the secretion rates normally found in mid-term development (Wolswinkel and Ammerlaan, 1986). Wolswinkel (1987) has also studied soybean seeds at stages younger than those selected by Thorne and his colleagues. The rates of secretion of sugar and amino compounds were increased over periods of 4 h by an external solution containing 250 mM mannitol compared to one containing only 50 mM mannitol. Thus young soybean seedcoats are similar in their behaviour to young pea and broad bean seedcoats.

This pattern seems appropriate for early stages of development, when solutes are first accumulating in the embryo sac vacuole, and are apparently secreted against concentration gradients (Section 6.1.3). In these stages, when the embryo is tiny, the seedcoats evidently possess the capacity to build up solute concentrations in the central vacuole of the embryo sac by increasing the rate of solute secretion as more solutes are accumulated. This period of apparently positive feedback corresponds to stages at which the seedcoats maintain their own pools of amino compounds, Pi and sugars around maximal concentrations (Figs 18, 21). The role of the tonoplast is no doubt also crucial in this period; its breakdown may signal an end to this phase of development.

As the embryo expands further and matures, the seedcoats meet its solute requirements by responding differently to external solute concentration. Here the pattern described for *Phaseolus vulgaris* (Patrick, 1984) and for pea at later stages (Wolswinkel et al., 1986) seems more appropriate - removal of sucrose from the apoplast by uptake into embryo cells promotes compensating secretion and further import by the seedcoats. The rate of sucrose utilization by the embryo, mediated by the sucrose synthase reaction, can now exert a dominant influence on the overall rate of sucrose import and transmission through the seedcoats. Marked genotypic differences are expressed among cultivars and between species as to the size of the internal sucrose pool(s), and the ultimate proportions incorporated into the various reserve materials of the embryo, including starch, other polysaccharides, and sucrose itself (Turner and Turner, 1957; Edwards and ap Rees, 1986a,b).

Maintenance of an optimal external sucrose concentration is critical for normal embryo development. The range of suitable values differs with species, as well as with stage of development. Pea embryos require an external concentration around 500 mM (= 18%, w/v). Without this high external concentration, a normal programme of storage protein synthesis that includes legumin cannot be undertaken, and extension of the radicle cannot be prevented (Stafford and Davies, 1979; Domoney et al., 1980).

7.5 Electrogenic ATPases and Cation Secretion

The acid pH of embryo sac liquid has already been noted (Section 6.1.3). The efflux of protons from seedcoat parenchyma is not restricted to the stoichiometric amounts required for co-transport with sucrose, glucose and amino compounds (Van Bel and Patrick, 1984, 1985). A surplus of protons is necessary to counteract the alkalinization of the external (apoplast) medium that would otherwise ensue from solute uptake by the developing embryo (Stafford and Davies, 1979). Proton co-transport is possibly more important for the uptake of amino compounds by the embryo than for sucrose uptake, where a major component of uptake is linear (non-saturable) (Lichtner and Spanswick, 1981; Thorne, 1982b; Bennett and Spanswick, 1983). The optimum pH for the uptake of [^{14}C]sucrose by developing bean cotyledons lies between pH 5.0 and pH 6.0 (Patrick, 1981). Furthermore, the molecular form of some amino compounds might be unsuitable for carrier-mediated uptake if the apoplast were not acid. Changing the pH of the solution circulating in pea seedcoat cups from pH 5.5 to pH 8.1 accelerates the rate of solute secretion, and the reverse is also true (Minchin and McNaughton, 1986). This confirms that regulation of external pH is itself an important component of the control of solute secretion (Section 6.3.2).

An electrogenic ATPase from the plasmalemma fraction isolated from pea embryos has been described by Pollak and Mogensen (1986). It is the type that would promote an uptake of K^+ in exchange for protons in the roots of developing seedlings (Simon, 1984; Kennedy, 1986). This ATPase is not detectable in mature seedcoats, but could promote K^+/H^+ exchange if present in the plasmalemma of seedcoat parenchyma during development. However, seedcoats could not achieve a net secretion of both K^+ and H^+ if this were the only type of carrier operating.

The effects of [K^+] in solutions applied to the inner wall of the embryo sac appear to vary widely. Patrick (1984) has emphasized the independence of net K^+ efflux from bean seedcoats compared to the efflux of sucrose and amino compounds. Wolswinkel and Ammerlaan (1985b) observed that 25 mM or 50 mM K^+ enhanced the secretion of sucrose from young pea seedcoats (compared to zero external K^+) but the stimulatory effect on the release of amino compounds was much less. This behaviour of K^+ as a solute may be non-specific.

A strong presumption remains that the critical control over carrier-mediated secretion of solutes into the embryo sac resides with the plasmalemma of the secretory cells

(Fig. 33). The observed stimulation of sucrose secretion by fusicoccin (Table 22) is consistent with the operation of electrogenic ATPases in the plasmalemma of secretory seedcoat parenchyma, rather than with a phloem unloading process that occurs strictly through the plasmodesmata. There is an obvious need for detailed studies on the transporters from the plasmalemma and tonoplast of seedcoat parenchyma, and their changing abundance and capacities during seed development.

7.6 Seedcoat Vascular Function and Integration of Seed Development

Developing seeds are much richer sources of endogenous growth substances than the fruit tissues surrounding them. The idea that fruit growth and development are regulated by growth substances emanating from seeds has gained much experimental support (Leopold and Kriedemann, 1975; Salisbury and Ross, 1978, 1985). There are strong correlations between relative growth rates of fruit parts, and profiles of growth substances in the attached seeds. For example, Eewens and Schwabe (1975) observed that the most rapid elongation of the pea pod occurred while gibberellins and then auxins increased to maximal concentrations in the embryo sac liquid. They further demonstrated that an applied mixture of GA_3 and NAA was most effective in promoting the growth of pods in which the seeds had been killed by piercing.

We are left with the question whether endogenous growth substances from embryo sac liquid originate in the embryonic suspensor (Section 6.1.1), or are secreted by the seedcoats. Recent evidence supports the view that the seedcoats supply growth substances inwardly to the embryo sac, and outwardly to the attached organs of the maternal plant.

In developing soybean seeds, the seedcoats contain most of the IAA present in any part of the seed or fruit (Hein et al., 1984). Exogenous IAA supplied to the embryo sac can accelerate secretion of sucrose, and sucrose import from the phloem (Table 22). Endogenous auxins are likely to be involved in proton pumping related to cell-wall 'loosening' (Salisbury and Ross, 1978,1985), and the most rapid rate of net water uptake in developing broad bean seeds is correlated with maximal seed content of 4-Cl-IAA (Pless et al., 1984). At least in the tribe Vicieae, this endogenous auxin is implicated in the control of cell expansion with concomitant water and solute acquisition by developing embryos. Water uptake by the seedcoats, embryo sac vacuole and embryo must be closely regulated; ion pumps are likely to regulate water fluxes as well as solute fluxes.

The seedcoats are able to import an amount of water surplus to the whole seed's requirements throughout development (until abscission from the funiculus, if this occurs) as a consequence of gaining solutes from the phloem. Water accompanying these solutes is also drawn from the phloem. A measurable net loss of water from the seed, beginning first in the seedcoats, then in the embryo, certainly characterizes the mid to final stages of seed development in peas (Murray, 1979b). Jenner (1982) has provided similar measurements for developing wheat grains.

How then does a developing seed return water to the maternal plant? Given the relative impermeability of the epidermal cuticle surrounding the testa, and disregarding the micropyle, since it is often blocked, evacuation of water from attached legume seeds must occur via the funiculus and the hull (Section 5.2.2), otherwise incoming phloem sap would be diluted. Within the seedcoats, water should be returned through xylem in response to pressure alone, provided the apoplast adjacent to xylem vessels is not strongly enriched in solutes. This proviso is easily reconciled with the extremely narrow distribution of xylem compared to phloem within the seedcoat vasculature. In soybean seedcoats, for example, most of the anastomosing veins consist entirely of phloem (Thorne, 1981). In the numerically simpler systems of pea and broad bean seedcoats, xylem is confined to a single longitudinal vein (Hardham, 1976; Offler and Patrick, 1986).

This apparent need to keep the apoplast clear of solutes near the xylem is another good reason to support the proposal that secretion of solutes through the plasmalemma to the apoplast leading to the embryo sac is the responsibility of restricted populations of parenchyma cells, after translocated solutes or their metabolic derivatives have been transferred to these cells via symplastic pathways (Fig. 33).

Reverse flow of water via xylem affords the mechanism for transmission of growth substances from the seedcoats to other parts of the maternal plant. Noodén (1984) has suggested that senescence signals (as yet unidentified) must move from maturing seeds to maternal plant parts, in order to promote senescence of these parts, and consequent mobilization of reserves in favour of the seeds. This applies to 'monocarpic' senescence, and not necessarily to plants that are not annuals. In soybean, either IAA or a precursor derived from [^{14}C]tryptophan injected into the developing seed can be transported from the seedcoats to other parts of the fruit and subtending leaves (Hein et al., 1986). This does not necessarily equate IAA with the senescence signal, but demonstrates that the transfer of growth substances from seedcoats to leaves is feasible.

The final period of desiccation undertaken by developing seeds is not separated temporally from the accumulation of reserves in the embryo. Up to half the protein content of the embryo can be acquired while there is an absolute loss of water from the seed. There is clearly an increased secretion of nitrogenous solutes relative to sucrose, and in relation to the complex feedback regulating phloem import (Fig. 33), it is interesting that the measured capacity of maturing soybean cotyledons to take up sucrose declines more rapidly than their capacity to take up glutamine (Ver Nooy et al., 1986). Thus nitrogen mobilized from senescing plant parts is preferred over sucrose when the maturing seeds become the only functional 'sink' for reception of solutes from phloem.

CHAPTER 8
Construction of Media for Embryo Growth *in vitro*

It is often desirable to grow isolated embryos or their component organs *in vitro*, or to generate embryoids from protoplasts, cells or tissues in culture (somatic embryogenesis), for the reasons discussed in Chapter 9. This chapter takes up the concern expressed by Norstog and Smith (1963) and reiterated by Ryczkowski (Section 6.1.3) that media developed for the culture of embryos should adequately reflect the authentic nutrition of the embryo *in vivo*, received while enclosed within seed and fruit coats and attached to the maternal plant.

Raghavan (1977) has compiled a list of 22 basal mineral media commonly adopted for embryo culture, most of which were originally devised for the culture of root tips (White, 1943; Bonner, 1943; Street, 1954). Their high NO_3^- contents may be convenient for preparing solutions, and appropriate for root culture, but NO_3^- is not an authentic source of nitrogen for embryos developing *in vivo* (Section 8.1). Furthermore, the concentration of sucrose is likely to be too low, since concentrations above 4% (w/v) inhibit root growth (Street and Lowe, 1950).

The obvious problems for developing embryos placed in many of these standard media are: high concentrations of nitrate; high Na^+ to K^+ ratios; low sucrose concentration; the lack of a source of sulphur other than sulphate; insufficient Pi for prolonged culture; non-optimal pH. With rare exceptions, embryo culture is characterized by unadventurous adherence to inappropriate media. Guidelines for reformulating media can now be framed according to what happens during seed development in the intact plant.

8.1 Preferred Nitrogen Sources for Embryo Growth *in vitro*

8.1.1 Nitrate is Inadequate

It has long been known that nitrate as sole source of nitrogen does not permit the rate of embryo growth *in vitro* to match the usual rate *in vivo*. For cotton embryos (*Gossypium hirsutum*), Mauney (1961) used casein hydrolysate and coconut 'milk' (embryo sac liquid) to supplement a nitrate-rich basal medium. Coconut milk is ideal for inclusion in media as it is already sterile, although the amino acid composition varies with stage of development (Baptist, 1963; Raghavan, 1977). For barley embryos, Norstog and Smith (1963) developed a medium with alternative nitrogen sources - glutamine plus selected amino acids, all combinations of which were superior to nitrate alone (see Fig. 34).

Nitrate is not only ineffective, but also unphysiological. Nitrate is not normally taken up into the phloem, the sole vascular supply route for seedcoats, and is not made available to the embryo sac. This was shown clearly by Millerd et al. (1975). Cotyledons removed from developing pea seeds 10 days after flowering exhibited no detectable nitrate reductase activity when extracted, but if excised cotyledons were cultured for two days in the presence of KNO_3, then a slight nitrate reductase activity was induced in them. Similar results were obtained by Lea et al. (1979). Nitrate reductase activity would have been detectable in cotyledon extracts immediately if the developing embryos had been exposed to nitrate *in vivo*.

8.1.2 Glutamine versus Asparagine

The importance of glutamine as a preferred source of combined nitrogen for developing embryos of *Capsella bursa-pastoris* and other species was established by Rijven (1952, 1956). Glutamine is likely to be a major source of nitrogen for developing cereal grains because of its high representation in prolamin and glutelin storage proteins in the endosperm (see Dalling and Bhalla, 1984). Embryos from cereal grains also rely heavily on glutamine, as demonstrated by Norstog and Smith (1963), who cultured barley embryos (Table 23). Of the amino compounds tested individually, best growth was obtained with glutamine (Fig. 34). The combination of glutamine with alanine was even better, and best results were obtained when this combination was supplemented with small quantities of five more amino acids, including the sulphur-amino acid cysteine (Fig. 34, Table 23).

FIG. 34. Maximum length attained in 10 days by barley embryos *in vivo* (A) or following transfer at the 0.4 mm stage to basal (White) medium plus 20% coconut milk (B), basal medium alone (C), basal medium plus alanine (400 mg L^{-1}) (D), basal medium plus glutamine (400 mg L^{-1}; E), basal plus both (F) or on complete medium (G: Table 23).

Table 23.

The Composition of the Medium Devised by Norstog & Smith (1963) for *in vitro* Culture of Barley Embryos.

Constituent	mg L^{-1}	Constituent	g L^{-1}
$MgSO_4.7H_2O$	730	Sucrose	90
$Ca(NO_3)_2.4H_2O$	290	Agar	9
Na_2SO_4	200		mg L^{-1}
KNO_3	160	Malic acid	100
KCl	140	Glutamine	400
$NaH_2PO_4.H_2O$	800	Alanine	400
Fe.citrate	10	Leucine	20
$MnSO_4.H_2O$	3	Tyrosine	20
$ZnSO_4.7H_2O$	0.5	Phenylalanine	10
H_3BO_3	0.5	Cysteine	10
$CuSO_4.5H_2O$	0.25	Tryptophan	10
$Na_2MoO_4.2H_2O$	0.25	Niacin	1.25
$CoCl_2.6H_2O$	0.25	Thiamin	0.25
		Pyridoxine	0.25
		Ca.pantothenate	0.25

The prominence of glutamine as a nitrogen source for developing cereal grains has been confirmed in studies of the composition of phloem sap from wheat ears (Fisher, 1987) and of glutamine delivery and metabolism in developing grains of wheat (Fisher and Gifford, 1986) and maize (Misra and Oaks, 1985). Glutamine has been incorporated into media used for regenerating maize plants from protoplasts (Qi-gui et al., 1987). It is difficult to justify the seemingly arbitrary choice of asparagine as the major or sole source of combined nitrogen for the culture of cereal embryos (Gamborg et al., 1977; Larkin et al., 1984) or endosperm cells (Felker, 1987). For cereal grains developing *in vivo*, asparagine is just one of a number of subsidiary nitrogen sources supplied together with glutamine; it is never a predominant source.

The assumption that asparagine is the major nitrogenous solute supporting protein synthesis in legume embryos developing *in vivo* (Atkins et al., 1975) is also incorrect (Section 6.3.2). This assumption gained credence following the report of Millerd et al. (1975) that asparagine was the most effective individual compound replacing half the usual amount of an amino acid mixture in a modified Linsmaier-Skoog (1965) medium for the culture of excised developing pea cotyledons. This comparison is invalid, however, because the media were autoclaved following the addition of amino compounds (Section 8.9), thereby converting glutamine to ammonium plus glutamate. Neither ammonium nor urea supports protein synthesis in cultured pea cotyledons (Table 24).

Thompson et al. (1977) employed glutamine successfully for the culture of excised soybean cotyledons, and found glutamine to be superior to asparagine or an amino acid mixture. Lea et al. (1979) subsequently showed that asparagine and glutamine were about equally effective as sole source of nitrogen in supporting growth and protein synthesis of excised pea cotyledons (Table 24).

However, *in vivo* neither glutamine nor asparagine is ever the sole source of nitrogen. Asparaginase activity in the developing pea embryo may restrict the usefulness of asparagine, since measured asparaginase activities (Fig. 24) are insufficient to account fully for observed rates of accumulation of amino groups *in vivo* (Murray and Kennedy, 1980). That the best results for the growth of pea embryos *in vitro* are obtained with glutamine plus asparagine reflects the fact that some combination of amides and

Table 24.

Net Increase or Decrease in the Protein Content of Excised Pea Cotyledons Incubated for 5 Days in Culture with Various Compounds Supplied as Sole Source of Nitrogen.

Added nitrogen source	Protein increase (mg/cotyledon)	Increment due to nitrogen source
None	3.04	zero
Glutamine	9.59	6.55
Asparagine	9.51	6.47
Glutamate	5.39	2.35
Nitrate	4.92	1.88
Allantoin	4.68	1.64
Ammonium	2.42	-0.62
Urea	2.86	-0.18

Data of Lea, Hughes and Miflin (1979).

* * * * * *

amino acids is secreted by the seedcoats (Chapter 6).

As well as serving as a major nitrogen source throughout development, glutamine may have morphogenetic effects necessary for embryogenesis. In a study of somatic embryogenesis from cells of *Gossypium klotzschianum* in suspension culture, glutamine was found to promote the differentiation of embryoids at an optimal concentration of 10-15 mM (Price and Smith, 1979). Asparagine had only a weak (infrequent) morphogenetic effect, at an optimal concentration of 5 mM. Similar results were obtained by Wetherell and Dougall (1976) for embryogenesis from suspension cultures of wild carrot. The morphogenetic effects of glutamine, and of very few other amino compounds such as alanine resemble the effect of NH_4^+ (Section 8.1.3). Perhaps it is more than coincidental that both glutamine and alanine are major secreted products of asparagine metabolism in the seedcoats of young pea seeds (Figs 25-27). This combination of solutes is clearly important for embryogenesis in many species (Fig. 34; Rijven, 1952).

8.1.3 Ammonium

Although NH_4^+ cannot be utilized as sole source of nitrogen for protein synthesis and growth of legume embryos, there is evidence that NH_4^+ promotes embryogenesis in cell cultures of some species, such as wild carrot (Halperin and Wetherell, 1965). As little as 5 mM is effective for this species. Stewart and Hsu (1977) found that NH_4^+ was most effective for the culture of cotton embryos *in ovulo* at an optimal concentration of 15 mM, and allowed embryos to develop to maturity.

The addition of NH_4^+ to a nitrate-rich medium facilitates a limited utilization of nitrate by embryos of some species, an effect also demonstrated for cells in suspension culture (Gamborg and Shyluk, 1970; Bayley et al., 1972). It is still not certain whether NH_4^+ is acting *per se* to bring about all of these effects, since substitution by glutamine and a few other amino compounds has been documented (Section 8.1.2). It is clear, however, that exogenous NH_4^+ elicits morphogenetic and developmental effects that cannot be reproduced by endogenous NH_4^+ released by the action of asparaginase on asparagine. In contrast to nitrate, NH_4^+ is present in seedcoat secretions (Section 6.1.3), and should be included in media used for embryo culture.

8.2 Sources of Sulphur

The quantitative distribution of reserve proteins in legume embryos can be influenced by inadequate sulphur supply even if the total amount of protein synthesized remains unchanged. Whether the total amount is diminished or not depends on the severity of the deficiency. In broad bean embryos, the cysteine content of the proteins is much more seriously diminished by sulphur-deficient conditions than the methionine content (Eppendorfer, 1971). In legume embryos generally, the representation of proteins enriched in sulphur-amino acids is minimized, whereas the proportion of proteins with zero or very low sulphur-amino acid content is maximized (Blagrove et al., 1976; Randall et al., 1979; Beach et al., 1985).

Methionine has been identified in the seedcoat secretion from pea, broad bean and cowpea (Section 6.3.2) but it is not yet certain whether any other sulphur compounds are also transmitted by seedcoats. Other possibilities include the tripeptides glutathione or homoglutathione received by seedcoat cells from phloem sap (Section 4.4.2), cysteine derived from the hydrolysis of these tripeptides in seedcoats, and sulphate derived from phloem sap. The concentrations of these other sulphur-metabolites in seedcoat secretions need to be determined, and the changes in sulphur supply during seed development need to be documented. It is not known, for instance, whether the glutathione stored in cotyledons (Spragg and Yemm, 1954; Macnicol and Randall, 1987) is translocated intact from phloem, or resynthesized *in situ*.

Studies of legume embryos cultured *in vitro* have nevertheless confirmed that the methionine concentration in the culture medium strongly influences the nature and relative amounts of reserve proteins synthesized in the embryo. Soybean embryos grown in medium supplemented with methionine exhibit a higher ratio of glycinin (11S protein) to conglycinins (7S proteins) (Table 25). Without the methionine supplement, soybean embryos are unable to synthesize sufficient methionine to maintain normal rates of storage protein synthesis. The growth-promoting effects of methionine cannot be reproduced by cysteine, aspartate, threonine, valine, isoleucine, leucine, proline, tyrosine, phenylalanine, histidine, lysine or arginine (Thompson et al., 1981).

Smith (1973) found that methionine was present at a concentration of 6.5 mM in embryo sac liquid from *Phaseolus*

Table 25.

Fresh Weight, Total Protein, Glycinin and Conglycinin Contents of Developing Soybean Cotyledons Incubated With or Without Supplementary Methionine (8.4 mM).

Time in culture (days)	Met +/-	Cotyledon Fresh weight (mg)	Total protein mg g^{-1}	Ratio 11S to 7S	11S + 7S as % total protein
3	-	71.3	61.1	0.9	39.7
3	+	72.5	57.5	1.3	44.7
5	-	97.5	57.5	0.8	44.3
5	+	106.3	78.8	1.1	44.5
7	-	129.1	55.2	0.8	57.5
7	+	145.3	71.8	1.8	55.5

Data of Holowach, Thompson and Madison (1984a).

* * * * * *

vulgaris. This is similar to the optimal concentration of methionine for the culture of soybean embryos, which is 8.4 mM (Thompson et al., 1981; Holowach et al., 1984a,b). However, Barratt (1984a) found that methionine in the range 1-5 mM inhibited growth and protein accumulation of broad bean embryos; the optimal concentration for this species is only 0.5 mM, in keeping with the low concentration of methionine secreted by seedcoats of broad bean and pea (Wolswinkel and de Ruiter, 1985; Table 19).

For embryos of both soybean and broad bean, sulphate cannot substitute for methionine, and at concentrations of 7.5 mM or greater, inhibits growth and protein synthesis (Barratt, 1984b; Holowach et al., 1984b). This concentration of sulphate could readily be approached or exceeded if the basal mineral medium were modified to reduce phosphate or nitrate contents (Table 23). Some compromise in basal mineral medium composition is therefore necessary to avoid inhibition by sulphate and ensure a suitable growth rate.

Although the contents of sulphur-amino acids in legume seed storage protein are of major importance for human and animal nutrition (Murray, 1979a, 1984a,b; Sastry and Murray, 1987), high contents of sulphur-amino acids are not essential for the successful germination of the embryo and establishment of the seedling. This is demonstrated by species and cultivars with genetically determined low sulphur amino acid content in the embryo, such as cowpea and mung bean. Such seedlings develop normally after germination even without an external supply of sulphate (Krishna and Murray, 1988).

8.3 Sugars

Sucrose is generally preferred to hexose as the major carbon and energy source for *in vitro* culture of embryos, and embryonic axes (Davis, 1983a,b). The concentration of sucrose necessary to maintain development (as distinct from growth following germination) is critical, and varies with species and stage of development (Section 7.4). Another possible effect of sucrose, not given by glucose or fructose, is suppression of chlorophyll synthesis (Edelman and Hanson, 1971).

Since seedcoats secrete a mixture of sucrose and glucose with negligible fructose (Section 6.3.3), it appears likely that glucose has some particular role to play in embryo development and nutrition. This is confirmed by Power (1987), who observed an altered morphology of shoots generated from slices of developing sunflower cotyledons in response to glucose, given either as sole sugar, or in combination with sucrose. Glucose appeared to improve regeneration, and stimulated the production of denser, longer epidermal hairs on the newly differentiated shoots. Glucose has no discernible effect on the growth of embryonic axes excised from imbibed pea embryos (Davis, 1983b). Observations on other species are needed before general conclusions can be drawn about the advantages of providing both sucrose and glucose for embryo culture.

8.4 Phosphate

The standard mineral media listed by Raghavan (1977) contain insufficient phosphate for prolonged culture of embryos. The addition of extra phosphate has often been recommended (Norstog and Smith, 1963; Murashige, 1974; Millerd et al., 1975). A concentration of 5 mM Pi is satisfactory, and consistent with the range of concentrations found in embryo sac liquid from pea seeds (Section 6.1.3). However, it is still not possible to prepare a medium that will provide enough Pi for prolonged embryo growth *in vitro*. Too high an initial concentration will lead to the formation of precipitates with other medium constituents, especially divalent cations. It is preferable to adopt a recipe which produces no precipitate, and to move embryos to fresh medium periodically as they develop (Stafford, 1978). The strategy described by Murashige (1974) for embryogenesis from callus culture involves at least two such transfers: one to promote shoot development, than another to promote root development (Section 8.8).

8.5 Organic Acids

Norstog and Smith (1963) included malic acid in their medium (Table 23) and suggested that it served as an effective buffer. Mauney et al. (1967) confirmed that malate is the principal organic acid anion present in embryo sac liquid from developing cotton seeds, and showed that malate is superior to citrate or succinate in media supporting the growth of cotton embryos. Smith (1973) measured malate concentrations around 20 mM in embryo sac liquid from developing bean seeds. Malate is an authentic secreted product of seedcoat metabolism (Fig. 25) and functions as a chelating agent for mineral cations as well as a buffer. Its inclusion in media is strongly recommended.

8.6 Optimum pH

The optimal pH varies with species, within a fairly narrow range. Norstog and Smith (1963) found that pH 4.9 was optimal for the growth and differentiation of barley embryos, whereas pH 5.6 or higher values were distinctly sub-optimal. For *Daucus carota* cells, embryogenesis and embryo growth both show optima around pH 5.4, with growth more tolerant of minor shifts than embryogenesis (Wetherell and Dougall, 1976). The range of pH values commonly employed, 4.9 to 5.5, coincides with values observed for embryo sac liquid (Section 6.1.3) and to optimal pH values observed for the uptake of sucrose (Patrick, 1981) and glutamine (Van Bel and Hermans, 1977).

The maintenance of optimal pH during culture is important, but presents practical difficulties. Although Millerd et al. (1975) claimed that excised pea cotyledons acidify their culture medium, Thompson et al. (1977) and Stafford (1978) observed that cultured legume cotyledons render an acid medium more alkaline. Stafford (1978) found that Pi is an effective buffer, but preferred to omit extra Pi in order to avoid precipitation in the medium, which would have prevented filter-sterilization. She maintained pH by transferring embryos to fresh medium every 2 days.

In critical comparative experiments, Wetherell and Dougall (1976) readjusted pH to the initial value of 5.4 every 8 hours, using 0.005 M NaOH or HCl according to the nitrogen sources in the media. Utilization of NH_4^+ as sole nitrogen source by cultured cells commonly leads to acidification of the external medium (Murray, 1969), whereas utilization of nitrate leads to alkalinization (Kennedy, 1986). For routine culture such exacting procedures are not necessary, especially if malate is included (Section 8.5), and Na^+ is replaced with K^+ (Wetherell and Dougall, 1976).

8.7 Vitamins

The addition of assorted vitamins such as niacin (nicotinic acid), thiamin, pyridoxine and pantothenate to nutrient media for embryo culture can be traced to Van Overbeek et al. (1942). Of these cofactors, only thiamin and pantothenate are present in substantial concentration in coconut milk (Vandenbelt, 1945). Today the inclusion of vitamins in media used for embryo culture is more often based on habit than on any demonstrated requirement.

Rigorous scientific demonstrations that vitamins are needed by embryos in culture are lacking. In fact, evidence that added vitamins are unnecessary has been mentioned by many authors. For example, Norstog and Smith (1963) included four vitamins in their medium (Table 23), but stated that "some of these were ineffective". Beasley and Ting (1973) tested whether pyridoxine, thiamin, niacin and *myo*-inositol were needed by cotton embryos, and concluded that "requirements for these organic constituents could not be established". Millerd et al. (1975) found that inositol and thiamin were unnecessary for the culture of pea cotyledons.

It is well known that legume seeds and nuts of various types are rich sources of thiamin, niacin and folate (e.g. Thomas and Corden, 1970; Wills et al., 1984). During seed development, the embryos of many species synthesize these and other cofactors from simple precursors (e.g. pea, Spragg and Yemm, 1954; soybean, Molin and Fites, 1980). The synthesis of *myo*-inositol is an integral part of the storage of phytin in the embryo (Lott, 1984). The ability to synthesize *myo*-inositol and a full range of nitrogenous cofactors is widespread among embryos of diverse plant species. There is no need to add any of these cofactors to culture media routinely unless a specific requirement can first be demonstrated for a given cultivar. Pantothenate is required for a particular cell line from *Datura innoxia*

(Savage et al., 1979), but such auxotrophic mutants are rare (Shimamoto and Nelson, 1981).

8.8 Growth Substances

Embryogenesis from callus culture will often not take place unless the exogenous auxin:cytokinin ratio is lowered (Skoog and Miller, 1957; Murashige, 1974). For example, a medium containing 5 μM 2,4-dichlorophenoxyacetic acid and 10-50 μM zeatin is suitable for callus production and shoot formation from immature embryos of *Sorghum bicolor* (Gamborg et al., 1977). Once shoots have formed, the auxin concentration can be increased again, otherwise root formation may remain suppressed by cytokinins. Root formation can also be promoted by the addition of ABA instead of IAA (Tsai and Tseng, 1979; Janardhanan and Lakshmanan, 1982a, 1982b; Henderson et al., 1984). It is important to assess the effectiveness of each of the various natural and synthetic growth substances that are commercially available, in view of wide species differences in response (Murashige 1974).

The effects of different exogenous growth substances on legume embryos taken into culture from developing seeds sometimes include altered patterns of dry matter and protein accumulation, and changes in protein composition, as in *Phaseolus vulgaris* (Sussex and Dale, 1979) and soybean (Ackerson, 1984). Abscisic acid can alter the proportions of the different storage proteins in these species, and in pea (Schroeder, 1984). In contrast, NAA or kinetin (10^{-5} M each) lower the rates of increase in fresh weight and dry matter contents of pea embryos (Stafford, 1978).

Whether endogenous growth substances have similar effects on embryos developing *in vivo* cannot be answered convincingly from such results *in vitro*. Embryos removed from developing legume seeds can complete their development *in vitro* without the addition of any growth substances to the medium. Embryogenesis and growth of barley embryos is also possible in the complete absence of added growth substances (Table 23, Fig. 34).

If endogenous growth substances act *in vivo*, are they produced within the embryo itself, or in the seedcoats (Section 7.6)? Do seedcoats secrete particular classes of growth substances in peaks at distinct stages of development, or uniformly throughout development, or not at all? These are relevant physiological questions that await investigation. Until profiles of growth substances have been monitored in seedcoat secretions obtained over a complete spectrum of developmental stages for at least one species, and complementary profiles have been recorded for the embryo as well, firm conclusions about the precise regulatory functions of growth substances in embryo development cannot be drawn.

8.9 Sterilization Procedures

A 'defined' medium is the stated aim of many researchers. In practice, most media fall short of this ideal because they have been autoclaved. A medium is no longer defined if autoclaving leads to substantial breakdown of the intended substrates, or causes significant production of unknown compounds with their own potential to inhibit or promote growth or morphogenesis.

Incorrect conclusions about the utilization of glutamine by pea embryos arose from the susceptibility of glutamine to hydrolysis on autoclaving (Section 8.1.2). Rijven (1952) drew attention to this problem long ago, and recommended that media containing glutamine should be filtered. Straus and La Rue (1954) observed that Seitz-filtered yeast extract in a sucrose-mineral medium was almost twice as effective as autoclaved yeast extract in supporting the growth of maize endosperm. The growth of cotton embryos is stimulated more by filtered coconut milk compared to coconut milk autoclaved in the medium (Mauney, 1961). All complex organic nitrogen sources such as glutamine, asparagine, arginine, tryptophan and urea should be filter-sterilized (Wetherell and Dougall, 1976), and so too should growth substances (Beasley and Ting, 1973).

Different sugars are usually autoclaved with the balance of the medium, or sometimes separately (Ferguson et al., 1958). Using ^{14}C-labelled sucrose, Ackerson (1984) has confirmed that sucrose can be autoclaved without detectable breakdown. Fructose, however, has sometimes given odd effects after autoclaving, and in critical comparisons it should be filter-sterilized (Verma and Dougall, 1977; Davis, 1983a).

8.10 Antibiotics

Antibiotics such as D-*threo*-chloramphenicol, penicillin, and streptomycin are sometimes added to culture media for tissue explants as a precaution against bacterial contamination (e.g. at 50 μg mL^{-1}; Davis, 1983a). In addition to this function, some antibiotics may themselves promote bud or root initiation. These include kanamycin (Owens, 1979a) chloramphenicol, erythromycin, benzylpenicillin, roscillin streptomycin and doxycycline (Lakshmanan and Janardhanan, 1983). All of these antibiotics can promote the differentiation of roots in hypocotyl explants from legume seedlings: *Lablab purpureus*(Raju et al., 1983), *Vigna radiata* (Shanmugasundaram et al., 1983) and *V. mungo* (Kumar et al. 1983). The lowest concentration effective in morphogenesis (0.1 μg mL^{-1}) yields results comparable to or superior to added IAA at 1-10 μg mL^{-1} (Section 8.8).

Adverse effects on protein synthesis at mitochondrial or chloroplast 70S ribosomes by certain of these antibiotics are unlikely at the low concentrations employed (Plummer et al., 1974; Wara-Aswapati and Bradbeer, 1974). Although their mechanisms of action are not yet understood, the choice of a suitable antibiotic may obviate the need for auxins in the final stage of somatic embryogenesis.

8.11 Conclusions

Media for the culture of embryos removed from developing seeds, or for somatic embryogenesis can now be formulated according to the composition of seedcoat secretions, rather than according to recipes originally devised for other purposes. The common inclusion of such ingredients as calf serum, casein hydrolysate, yeast extract and vitamins instead of defined organic sources of nitrogen smacks of witchcraft rather than science. The guidelines set out in this chapter will allow newcomers to this field to formulate media objectively, matching as nearly as possible the nutrition received by embryos developing *in vivo*.

CHAPTER 9

Embryogenesis and Prospects for Plant Improvement

9.1 Recognition of Somaclonal Variation

The review "Plant Propagation Through Tissue Culture" by T. Murashige (1974) provides a remarkably prophetic account of the aims and potential applications of tissue culture, most of which have been realized since. One of the original aims of somatic embryogenesis was the more rapid propagation of novel genotypes or virus-free stock. In principle, a single plant could give rise to a multitude of exact copies in a much shorter time through vegetative callus culture than through conventional breeding procedures depending on the sexual cycle.

The term 'embryoid' was introduced to distinguish non-zygotic embryonic plants derived from cells in culture (Haccius and Lakshmanan, 1969; Lakshmanan, 1969, 1975). More recently, the term 'somaclone' has been used to indicate both the non-zygotic origin of a regenerated plant, and the presumed clonal relationship of the originating cell(s), and the immediate progeny (Larkin and Scowcroft, 1981a,b). Instead of using 'R' to represent a regenerated plant, Larkin et al. (1984) have proposed that 'SC' represent each somaclone, then SC_1, SC_2, SC_3 and so on represent subsequent generations derived sexually by self-fertilization. However, Evans et al. (1984, 1987) prefer the neutral symbol 'R'. They suggest that the term 'somaclone' should be reserved for plants regenerated from somatic cells or protoplasts, and that 'gametoclone' be applied to plants derived from gametes or gametic tissue, as in anther culture (Zeng, 1983; Bollon and Raquin, 1987). Either system would maintain some analogy with the symbols F_1, F_2, F_3 . . . used for filial generations of plants obtained only through sexual cycles.

As well as providing the vehicle for regeneration of novel plants following attempted protoplast transformation (Mathews et al., 1985; Fraley et al., 1987; Tachini and Walbot, 1987), or somatic hybridization (protoplast fusion; Pelletier, 1987), the tissue culture process itself is now recognized as being strongly mutagenic. The word 'clone' is really a misnomer - and heritable variation amongst progeny that are no longer genetically identical is known as 'somaclonal variation'. A wide range of characteristics can be subject to somaclonal variation, including many of agronomic significance.

Among the most intensively studied species are sugar cane (Heinz and Mee, 1971; Heinz et al., 1977; Larkin and Scowcroft, 1981b, 1983a), maize (Gengenbach et al., 1977, 1981; Edallo et al., 1981), wheat (Larkin et al., 1984; Ryan and Scowcroft, 1987), rice (Oono, 1981; Zeng, 1983), tomato (Evans et al., 1984, 1987), lettuce (Engler and Grogan, 1984) and alfalfa (Reisch and Bingham, 1981; Pfeiffer and Bingham, 1984). Valuable reviews of this rapidly expanding field have been provided by Shepard (1981), Larkin and Scowcroft (1981a, 1983b), Scowcroft et al. (1983), Larkin et al. (1983, 1984), Larkin (1985, 1987) and Evans et al. (1984, 1987).

So abundant are somaclonal variants in culture that they may sometimes comprise the majority of regenerated plants. For example, among the selfed progeny of 1,121 rice somaclones, only 28.1% had not altered in at least one of five parameters scored (Oono, 1981). In view of the ease with which variation is induced, two questions seem particularly important. What is the molecular basis for the induction of stable mutations while cells are dividing and differentiating in culture? And how useful are somaclonal variants if they are unpredictably different from the parent?

9.2 Induction of Somaclonal Variation

9.2.1 Nutrient Stress

It is generally believed that some form of culture shock or stress may predispose the nuclear and organelle genomes of cultured cells to mutation (Austin and Cassells, 1983; Larkin and Scowcroft, 1983b). The frequency of variant somaclones often tends to increase with the duration of the preliminary period in culture (McCoy et al., 1982). Any one of a number of nutrient stresses (itemized in Chapter 8) might contribute to the phenomenon, e.g. the provision of asparagine instead of glutamine as major or sole source of combined nitrogen for callus culture derived from immature cereal embryos (Section 8.1.2).

Alternatively, the depletion of some key factor might be involved. Nutrient stress has not been investigated systematically to determine which nutrients, if any, are responsible for destabilising a previously stable genotype. Since conditions favouring stability can be chosen (Evans and Gamborg, 1982), the nutrient stress hypothesis is well worth further study.

9.2.2 Underlying Mechanisms

Evans et al. (1984, 1987) have suggested that some of the causes of somaclonal variation might include:

(a) gene deletion resulting from elimination of whole chromosomes, as in polyploids, or following the formation of intergeneric hybrids;

(b) intergenomic mitotic recombination (crossing over);

(c) reassortment of plastid genomes in novel combinations (the male gamete's plastid genome would normally be completely discarded on fertilization).

Larkin (1985) believes that visible chromosomal abnormalities are often a consequence of whatever causes somaclonal variation, rather than the primary cause. Since the apparent rates of mutation to dominant or co-dominant alleles not expressed in the parental genotype are so much higher than would normally occur, Larkin suggests that somaclonal variation probably involves transposable elements controlling gene organization and expression. The behaviour of transposable elements (the so-called 'jumping genes') is best understood in maize (McClintock, 1950; Fedoroff, 1983, 1984; Grandbastien, 1987). It is instructive to recall McClintock's subdivision of mutable gene loci in this species (1950, verbatim):

"The mutable loci fall into two major classes: (1) those that require a separate activator factor for instability to be expressed, and (2) those that are autonomous with respect to the factor that controls the onset of mutability. They also may be subdivided on a quite different basis. This is related to the types of expression of the mutations that occur. The following types are present: (a) changes from the mutant to, or close to, the wild-type expression. After such a mutation, the locus may be permanently stabilized. It may no longer show evidence of the instability phenomenon. (b) a second group, similar to (a) except that the mutation to wild-type does not produce stability of the locus. The wild-type-producing locus, in turn, may mutate to give the recessive expression. (c) a third type where the mutations give rise to a series of alleles of the affected loci. These alleles are distinguished by different degrees of quantitative expression of the normal phenotype. Most of these are relatively stable; only rarely does instability again appear. (d) a fourth type, similar to (c). Most of the alleles, however, are not stable, for they, in turn, can mutate in the direction of a higher or lower grade of quantitative expression of the phenotype. Mutable loci showing these different types of expression of mutation are found in both the major classes, that is, in the activator-requiring class, and in the autonomous class."

The parallels between this early assessment of mutability and the kinds of change documented for somaclonal variants are striking. In wheat, alterations from recessive to wild-type expression include the change from white to brown glume colour, and the synthesis of certain gliadins among endosperm storage proteins in the grain (Larkin et al., 1984). Transitions from the wild-type to the recessive expression, impossible to explain as a simple alteration to a single allele in a hexaploid, include the change from red to white grain coat colour (Larkin et al., 1984), and the appearance of male sterility and jointless pedicel in tomato (Evans et al., 1987).

Additive conditions, showing degrees of increase or decrease over uniform parental types, include plant height, seed yield (Oono, 1981; Reisch and Bingham, 1981), fruit pigmentation and fruit dry matter content (Evans et al., 1987).

In wheat, no chromosome loss has been observed in any somaclonal family, and similar variants do not occur among seed progeny of the parent. Non-segregating and segregating characteristics occur independently of one another (Larkin et al., 1984). For example, an individual somaclone family ('IB-91') can segregate for plant height, awn and α-amylase variants; not segregate for grain coat colour differences (red vs. white); and be invariant, as in the parent, for heading date and pale glume colour. Conversely, any particular characteristic could display all classes of variation - invariant, segregating and non-segregating - in different somaclone families.

Larkin (1985) has suggested a number of ways that a transposed element could affect gene expression in somaclonal variants:

(a) Disruptive insertion - the transposed element interrupts a structural gene, preventing its normal expression;

(b) Modified expression - a gene might be expressed at an enhanced level, or at a different stage in development, if a transposed element is placed in a promoter position ('upstream');

(c) Gene activation - silent members of multigene families could be expressed if a transposed element were excised and moved elsewhere;

(d) Piggy-backing - genes could be carried with transposable elements to new locations subject to different regulation;

(e) Concerted control - a single autonomous controlling element could regulate the expression of a number of non-autonomous elements and genes associated with them; the transposition of a non-autonomous element to a new location could subjugate a gene or genes at that site to new regulation;

(f) Intron insertion - an inserted transposable element could be transcribed with the gene and then treated as an intron, i.e. excised in the production of mature mRNA. If the excision is not exact, a slight frame shift in the translation of the mRNA could yield a functional protein with somewhat different properties. If this were a storage protein, such an alteration would usually not affect the function, which is to be degraded following imbibition and germination.

The best way to verify Larkin's proposals is to obtain DNA base sequence information for nuclear, mitochondrial and other plastid genomes from parental and variant regenerated plants. This can be done by treating replicate samples of isolated DNA with a battery of restriction endonucleases, each with different cleavage site specificity. The oligonucleotide fragments resulting from each digestion are separated by gel electrophoresis, and the fragment patterns compared. Differences in base sequence resulting from insertions, inversions or deletions will be detected as altered fragment patterns. This type of plastid genome analysis has been applied successfully to the determination of the phylogenetic affinities of maize and the various wild teosintes (Timothy et al., 1979; see Murray, 1984a), and to the characterization of male sterility associated with Texas (T) type cytoplasm, which confers susceptibility to the toxin produced by Southern corn leaf blight, *Helminthosporium (Drechslera) maydis*.

Male sterility is useful in female parents used for the breeding of conventional hybrids, but the prevalence of type T cytoplasm in maize cultivars grown in the U.S.A. led to a massive crop loss from blight in 1970 (Bradbeer, 1988). Somaclonal variants of maize resulting from culture in the presence or absence of the blight toxin have been found to be resistant to the toxin, and are male fertile (Gengenbach et al., 1977, 1981; Brettell et al., 1980; Umbeck and Gengenbach, 1983). Restriction endonuclease fragment analyses reveal that toxin resistance and male fertility are correlated with the loss of one particular fragment 6.6 kilobases long from the mitochondrial DNA (Brettell et al., 1982; Umbeck and Gengenbach, 1983), and with decreased synthesis of a mitochondrial polypeptide of M_r 13,000 (Dixon et al., 1982).

9.3 Application of Somaclonal Variation

Concern that many of the desirable characteristics of a parent cultivar might be lost during the genome destabilization accompanying the generation of somaclonal variants has been allayed by the recovery of these characteristics in the majority of progeny (Larkin and Scowcroft, 1983b). One selection pressure favouring such recovery is conservation of the capacity for embryogenesis in culture - cells so altered that they are no longer capable of mitotic division and embrogenesis are lost. A second selection pressure in favour of useful parental attributes is the rejection of regenerants that cannot form seeds - although this does not apply to sugar cane or potato.

Improvements over the parental type are readily recovered. Even though the process is unpredictable in detail, it is a thoroughly rewarding approach. One area of practical concern where beneficial results are being obtained is in the generation of somaclones resistant to specific pathogens. The recovery of maize variants with resistance to leaf blight toxin (Section 9.2) is just one instance of this. Others are listed in Table 26. Sometimes in the selection of somaclonal variants there is a bias towards toxin resistance even when the toxin is not included in the medium as an additional selective agent (Larkin and Scowcroft, 1983a).

The use of some pesticides with unwanted or unknown side-effects can be avoided if pathogen-resistant plants become the norm rather than the exception. It is fortunate too that 'character-stacking' (Shepard et al., 1980; Larkin and Scowcroft, 1983b) is feasible. This strategy entails the continued accumulation of desirable characteristics through new culture sequences and selection procedures as the need arises. In this way yield parameters can be maximized in pathogen-resistant cultivars. We can indeed witness evolution "*per saltum*" (by a jump) - a mechanism Charles Darwin put to one side (Eiseley, 1979).

In the short term, the selection of somaclones is far more likely to repay the time and money invested than attempts to transform plants by the introduction of foreign

Table 26.
Species for Which Somaclonal Variants with Toxin Tolerance or Toxin Resistance Have Been Selected.

Plant species	Toxin source	References
Sugar cane[a]	Fiji disease virus	Krishnamurthi &
	Downey mildew	Tlaskal, 1974
	Eyespot disease	Heinz et al., 1977
	(*Helminthosporium sacchari*)	Larkin & Scowcroft 1981b, 1983a
	Smut (*Ustilago scitaminia*)	Liu, 1981
Maize	Southern leaf blight	Section 9.2
	Early blight	Shepard, 1980,1981
	(*Alternaria solani*)	Shepard et al. 1980
	Late blight (*Phytophthora infestans*)	
Rapeseed	Blackleg (*Phoma lingam*)	Sacristan, 1982
Alfalfa	Wilt (*Fusarium oxysporum*)	Hartman et al., 1984a,b
Tomato	Wilt (*Fusarium oxysporum*)	Evans et al., 1984, 1987

[a] *Saccharum officinarum x S. spontaneum* hybrids

* * * * * *

DNA. Apart from the difficulties of first isolating and identifying a suitable gene, there is no guarantee that DNA introduced to a plant cell either via a plasmid or by electroporation will be expressed to an appropriate degree at the correct stage of the life cycle, if at all (Owens, 1979b).

Somaclonal variation takes advantage of the inbuilt reservoir of genomic variability that all eukaryotes possess. Destabilization of the genome by the tissue culture process allows direct selection for resistance to antibiotics (e.g. kanamycin, Owens, 1981) or to the toxins of pathogens (Table 26) or to some herbicides (Larkin, 1987). Gene transfer to the mitochondrial or chloroplast genomes from remote donors can be facilitated by somatic hybridization, and monitored by restriction endonuclease DNA-fragment analysis (Pelletier, 1987).

If the generation and selection of somaclones is already so effective, is there any urgent need to study the culture and nutrition of embryos removed from developing seeds? I believe there is, for the following purposes:

- to study nutrient uptake processes, and distinguish the metabolism of the embryo from the transformations of potential nutrients that take place in the seedcoats or other maternal supply organs;
- to study the effects of growth substances, and assist in the elucidation of their roles *in vivo*;
- to rescue embryos with genotypes that would be lost in seeds normally aborted, such as those resulting from infertile crosses.

This last application is particularly important for the production of new cultivars of seedless fruits, such as grapes (Emershad and Ramming, 1984) and banana (Krikorian and Cronauer, 1984).

References

Ackerson, R. C. (1984). J. Exp. Bot. **35**, 403-413.

Allen, S., Raven, J. A., and Thomas, G. E. (1985). J. Exp. Bot. **36**, 413-425.

Anderson, L. E., and Pacold, I. (1972). Plant Physiol. **49**, 393-397.

Andrew, C. S., and Pieters, W. H. J. (1972). 'Foliar Symptoms of Mineral Disorders in *Phaseolus atropurpureus*'. Division of Tropical Crops and Pastures Technical Paper No. **11**, CSIRO Australia.

Andrew, C. S., and Pieters, W. H. J. (1976a). 'Foliar Symptoms of Mineral Disorders in *Lotononis bainesii*'. Division of Tropical Crops and Pastures Technical Paper No. **17**, CSIRO Australia.

Andrew, C. S., and Pieters, W. H. J. (1976b). 'Foliar Symptoms of Mineral Disorders in *Glycine wightii*'. Division of Tropical Crops and Pastures Technical Paper No. **18**, CSIRO Australia.

Arnon, D. I. (1949). Plant Physiol. **24**, 1-15.

Arnon, D. I. (1958). In 'Trace Elements' (C. A. Lamb, O. G. Bentley and J. M. Beattie, Eds), pp. 1-32. Academic Press, New York.

Arnon, D. I., and Stout, P. R. (1939). Plant Physiol. **14**, 599-602.

Aston, M. J., and Jones, M. M. (1976). Planta **130**, 121-129.

Atkins, C. A., and Canvin, D. T. (1971). Can. J. Bot. **49**, 1225-1234.

Atkins, C. A., Pate, J. S., and Sharkey, P. J. (1975). Plant Physiol. **56**, 807-812.

Atkins, C. A., Pate, J. S., and Layzell, D. B. (1979). Plant Physiol. **64**, 1078-1082.

Atkins, C. A., Pate, J. S., Ritchie, A., and Peoples, M. B. (1982). Plant Physiol. **70**, 476-482.

Atkins, C. A., Pate, J. S., Peoples, M. B., and Joy, K. W. (1983). Plant Physiol. **71**, 841-848.

Atkins, C. A., Shelp, B. J., Storer, P. J., and Pate, J. S. (1984). Planta **162**, 327-333.

Atkins, C. A., Pate, J. S., and Peoples, M. B. (1986). In 'Fundamental Ecological and Agricultural Aspects of Nitrogen Metabolism in Higher Plants' (H. Lambers, J. J. Neeteson and I. Stulen, Eds), pp. 235-238. M. Nijhof, Dordrecht.

Attridge, T. H., and Smith, H. (1974). Biochim. Biophys. Acta **343**, 453-464.

Austin, S., and Cassells, A. C. (1983). Plant Sci. Lett. **31**, 107-114.

Babior, B. M. (1975). 'Cobalamin Biochemistry and Patho-Physiology'. Wiley Interscience. J. Wiley & Sons, N.Y.

Badger, M. R. (1985). Annu. Rev. Plant Physiol. **36**, 27-53.

Bain, J. M., and Mercer, F. V. (1966). Aust. J. Biol Sci. **19**, 49-67.

Baldy, P., Verin, C., and Cavalié, G. (1977). Abstracts 4th Int. Congress on Photosynthesis, pp. 15-16. U.K.

Baptist, N. G. (1963). J. Exp. Bot. **14**, 29-41.

Barlow, C. A., and McCully, M. R. (1972). Can. J. Zool. **50**, 1497-1498.

Barratt, D. H. P. (1984a). Ann. Bot. **54**, 21-30.

Barratt, D. H. P. (1984b). Ann. Bot. **54**, 31-38.

Bauer, A., Joy, K. W., and Urquhart, A. A. (1977a). Plant Physiol. **59**, 920-924.

Bauer, A., Urquhart, A. A., and Joy, K. W. (1977b). Plant Physiol. **59**, 915-919.

Bauer, W. D. (1981). Annu. Rev. Plant Physiol. **32**, 407-449

Bayley, J. M., King, J., and Gamborg, O. L. (1972). Planta **105**, 15-24.

Beach, L. R., Spencer, D., Randall, P. J., and Higgins, T. J. V. (1985). Nucleic Acids Res. **13**, 999-1013.

Beadle, N. C. W. (1952). Ecology **33**, 49-62.

Bean, R. C., and Todd, G. W. (1960). Plant Physiol. **35**, 425-429.

Bean, R. C., Porter, G. G., and Barr, B. K. (1963a). Plant Physiol. **38**, 280-284.

Bean, R. C., Porter, G. G., and Barr, B. K. (1963b).

Plant Physiol. **38**, 285-290.

Beasley, T. A., and Ting, I. P. (1973). Am. J. Bot. **60**, 130-139.

Becker, J. W., Reecke, G. N. Jr., Cunningham, B. A., and Edelman, G. M. (1976). Nature **259**, 406-409.

Bell, E. A. (1984). In 'Seed Physiology. Vol. 1. Development' (D. R. Murray, Ed.), pp. 245-264. Academic Press, Sydney.

Benner, J. L., and Noodén, L. D. (1984). Biochem. Physiol. Pflanzen **179**, 269-275.

Bennett, A. B., and Spanswick, R. M. (1983). Plant Physiol. **72**, 781-786.

Bennett, A. B., Sweger, B. L., and Spanswick, R. M. (1984) Plant Physiol. **74**, 434-436.

Bergersen, F. J. (1965). Aust. J. Biol. Sci. **18**, 1-9.

Bergersen, F. J. (1982). 'Root Nodules of Legumes: Structure and Functions'. Research Studies Press, John Wiley and Sons, Chichester U.K.

Bergfeld, R., and Schopfer, P. (1986). Ann. Bot. **57**, 25-33.

Berry, J. A., Osmond, C. B., and Lorimer, G. H. (1978). Plant Physiol. **62**, 954-967.

Besford, R. T. (1978). Ann. Bot. **42**, 317-324.

Bhalla, P. L., Singh, M. B., and Malik, C. P. (1979). Acta Bot. Indica **7**, 72-86.

Bhalla, P. L., Singh, M. B., and Malik, C. P. (1980). Biochem. Physiol. Pflanzen **175**, 263-267.

Bhalla, P. L., Singh, M. B., and Malik, C. P. (1981a). Z. Pflanzenphysiol. **103**, 115-119.

Bhalla, P. L., Singh, M. B., and Malik, C. P. (1981b). Biochem. Physiol. Pflanzen **176**, 789-792.

Biddulph, O. (1959). In 'Plant Physiology. Vol. II. Plants in Relation to Water and Solutes' (F. C. Steward, Ed.), pp. 553-603. Academic Press, New York and London.

Biddulph, O., and Cory, R. (1960). Plant Physiol. **35**, 689-695.

Biddulph, O., Nakayama, F. S., and Cory, R. (1961). Plant Physiol. **36**, 429-436.

Bidwell, R. G. S. (1974) 'Plant Physiology'. Collier MacMillan, London.

Bidwell, R. G. S. (1979) 'Plant Physiology'. 2nd edition. Collier MacMillan, London.

Bieleski, R. L. (1973). Annu. Rev. Plant Physiol. **24**, 225-

252.

Bils, R. F., and Howell, R. W. (1963). Crop Sci. **3**, 304-308.

Bird, I. F., Cornelius, M. J., Keys, A. J., and Whittingham, C. P. (1972). Phytochem. **11**, 1587-1594.

Bisson, C. S., and Jones, H. A. (1932). Plant Physiol. **7**, 91-105.

Blagrove, R. J., Gillespie, J. M., and Randall, P. J. (1976). Aust. J. Plant Physiol. **3**, 173-184.

Blomquist, R. V., and Kust, C. A. (1971). Crop Sci. **11**, 390-393.

Böhme, H. (1978). Eur. J. Biochem. **83**, 137-141.

Bollard, E. G. (1960). Annu. Rev. Plant Physiol. **11**, 141-166.

Bollard, E. G. (1970). In 'The Biochemistry of Fruits and Their Products' (A. C. Hulme, Ed.), pp. 387-425. Academic Press, London.

Bollon, H., and Raquin, C. (1987). Nestlé Research News 1986/7, 81-91.

Bonner, J. (1943). Bull. Torrey Bot. Club **70**, 184-189.

Bové, J. M., Bové, C., Whatley, F. R., and Arnon, D. I. Z. Naturforsch. Teil. B **18**, 683-688.

Bowes, G., Ogren, W. L., and Hageman, R. H. (1971). Biochem. Biophys. Res. Commun. **45**, 716-722.

Braaksma, F. J., and Feenstra, W. J. (1982). Physiol. Plant. **54**, 351-360.

Bradbeer, J. W. (1977). In 'The Molecular Biology of Plant Cells', Botanical Monographs Vol. **14** (H. Smith, Ed.), pp. 64-84. Blackwell, Oxford.

Bradbeer, J. W. (1988). 'Seed Dormancy and Germination' Blackie, Glasgow and London.

Brady, C. J. (1987). Annu. Rev. Plant Physiol. **38**, 155-178

Brettell, R. I. S., Thomas, E., and Ingram, D. S. (1980). Theor. Appl. Genet. **58**, 55-58.

Brettell, R. I. S., Conde, M., and Pring, D. R. (1982) Maize Genet. Coop. Newsletter **56**, 13-14.

Brink, R. A., and Cooper, D. C. (1947). Bot. Rev. **13**, 423-541.

Brown, P. H., Welch, R. M., and Cary, E. E. (1987). Plant Physiol. **85**, 801-803.

Brown, R. H. (1978). Crop Sci. **18**, 93-98.

Brownell, P. F. (1965). Plant Physiol. **40**, 460-468.

Brownell, P. F., and Crossland, C. J. (1972). Plant Physiol. **49**, 794-797.

Brownell, P. F., and Wood, J. G. (1957). Nature **179**, 635-636.

Bryce, J. H., and Wiskich, J. T. (1985). Aust. J. Plant Physiol. **12**, 229-239.

Butz, R. G., and Jackson, W. A. (1977). Phytochem. **16**, 409-417.

Cammack, R., Rao, K. K., and Hall, D. O. (1985). Physiol. Vég. **23**, 649-658.

Campbell, J. McA., and Wray, J. L. (1983). Phytochem. **22**, 2375-2382.

Campbell, L. C., Miller, M. H., and Loneragan, J. F. (1975). Aust. J. Plant Physiol. **2**, 481-487.

Campbell, W. H. (1978). Z. Pflanzenphysiol. **88**, 357-361.

Canny, M. J. (1973). New Phytol. **72**, 1269-1280.

Canvin, D. T., and Atkins, C. A. (1974). Planta **116**, 207-224.

Carroll, B. J., McNeil, D. L., and Gresshoff, P. M. (1985) Plant Physiol. **78**, 34-40.

Castignetti, D., and Smarelli, J. Jr. (1986). FEBS Lett. **209**, 147-151.

Chaillou, S., Morot-Gaudry, J.-F., Salsac, L., Lesaint, C. and Jolivet, E. (1986). Physiol. Vég. **24**, 679-687.

Cheniae, G. M. (1970). Annu. Rev. Plant Physiol. **21**, 467-498.

Christensen, T. M. I. E., and Jochimsen, B. U. (1983). Plant Physiol. **72**, 56-59.

Clandinin, M. T., and Cossins, E. A. (1972). Biochem. J. **128**, 29-40.

Clandinin, M. T., and Cossins, E. A. (1975). Phytochem. **14**, 387-391.

Clarkson, D. T., and Hanson, J. B. (1980). Annu. Rev. Plant Physiol. **31**, 239-298.

Claussen, W., Hawker, J. S., and Loveys, B. R. (1985a). J. Plant Physiol. **119**, 123-131.

Claussen, W., Loveys, B. R., and Hawker, J. S. (1985b). Physiol. Plant. **65**, 275-280.

Coker, G. T. III, and Schaefer, J. (1985). Plant Physiol. **77**, 129-135.

Collier, M. D., and Murray, D. R. (1977). Aust. J. Plant Physiol. **4**, 571-582.

Coombe, B. G. (1960). Plant Physiol. **35**, 241-250.

Cooper, D. C. (1938). Bot. Gaz. **100**, 123-132.

Copeland, L., Harrison, D. D., and Turner, J. F. (1978). Plant Physiol. **62**, 291-294.

Coughlan, M. (1980). 'Molybdenum and Molybdenum-Containing Enzymes'. Pergamon Press, Great Britain.

Crafts, A. S. (1961). 'Translocation in Plants'. Holt, Rinehart & Winston, New York.

Crawford, R. M. M. (1976). In 'Plant Structure, Function and Adaptation' (M. A. Hall, Ed.), pp. 197-253. The Macmillan Press Ltd, London.

Cronshaw, J. (1981). Annu. Rev. Plant Physiol. **32**, 465-484

Cruickshank, I. A. M., Dudman, W. F., Peoples, M. B., and Smith, M. M. (1987). Aust. J. Plant Physiol. **14**, 549-559.

Dalling, M. J., and Bhalla, P. L. (1984). In 'Seed Physiology. Vol. 2. Germination and Reserve Mobilization' (D. R. Murray, Ed.), pp. 163-199. Academic Press, Sydney.

Dalling, M. J., Tolbert, N. E., and Hageman, R. H. (1972a) Biochim. Biophys. Acta **283**, 505-512.

Dalling, M. J., Tolbert, N. E., and Hageman, R. H. (1972b) Biochim. Biophys. Acta **283**, 513-519.

Davies, D. D. (1973). In 'Biosynthesis and its Control in Plants' (B. V. Milborrow, Ed.), pp. 1-20. Academic Press, New York.

Davies, D. D. (1986). Physiol. Plant. **67**, 702-706.

Davies, M. E. (1964). Plant Physiol. **39**, 53-59.

Davis, B. D. (1983a). Am. J. Bot. **70**, 816-820.

Davis, B. D. (1983b). Am. J. Bot. **70**, 821-826.

Del Río, L. A. (1983). Life Chem. Rep. **2**, 1-34.

Del Río, L. A., Sevilla, F., Gomez, M., Yañez, J., and Lopez, J. (1978). Planta **140**, 221-225.

Del Río, L. A., Lyon, D. S., Olah, I., Glick, B., and Salin, M. (1983). Planta **158**, 216-224.

Delves, A. C., Higgins, A. V., and Gresshof, P. M. (1987). Aust. J. Plant Physiol. **14**, 689-694.

De Ruiter, H., and Kollöffel, C. (1983). Plant Physiol. **73**, 525-528.

De Ruiter, H., Schuurmans, J., and Kollöffel, C. (1984). J. Plant Physiol. **116**, 47-57.

Dickson, R. E., and Larson, P. R. (1975). Plant Physiol. **56**, 185-193.

Dilley, D. R. (1966). Plant Physiol. **41**, 214-220.

Dixon, L. K., Brettell, R. I. S., Gengenbach, B. G., and Leaver, C. J. (1982). Theor. Appl. Genet. **63**, 75-80.

Dixon, N. E., Gazzola, C., Blakely, R. L., and Zerner, B. (1975). J. Am. Chem. Soc. **97**, 4131-4133.

Dixon, N. E., Gazzola, C., Asher, C. J., Lee, D. S., Blakely, R. L., and Zerner, B. (1980a). Can. J. Biochem. **58**, 474-480.

Dixon, N. E., Hinds, J. A., Fihelly, A. K., Gazzola, C., Winzor, D. J., and Zerner, B. (1980b). Can. J. Biochem. **58**, 1323-1334.

Domoney, C., Davies, D. R., and Casey, R. (1980). Planta **149**, 454-460.

Donovan, G. R., Jenner, C. F., Lee, J. W., and Martin, P. (1983). Aust. J. Plant Physiol. **10**, 31-42.

Duke, M. V., and Salin, M. L. (1983). Phytochem. **22**, 2369-2373.

Dunnill, P. M., and Fowden, L. (1967). Phytochem. **6**, 1659-1663.

Eames, A. J., and MacDaniels, L. H. (1947). 'An Introduction to Plant Anatomy'. McGraw-Hill, N. Y. and London.

Edallo, S., Zucchinau, C., Perinzin, M., and Salamini, F. (1981). Maydica **26**, 39-56.

Edelman, J., and Hanson, A. D. (1971). Planta **98**, 150-156.

Edwards, G. E., and Walker, D. A. (1983). 'C_3,C_4:Mechanisms, and Cellular and Environmental Regulation of Photosynthesis'. University of California Press,U.S.A.

Edwards, J., and apRees, T. (1986a). Phytochem. **25**, 2027-2032.

Edwards, J., and ap Rees, T. (1986b). Phytochem. **25**, 2033-2039.

Edwards, J., ap Rees, T., Wilson, P. M., and Morrell, S. (1984). Planta **162**, 188-191.

Ehleringer, J. R. (1978). Oecologia **31**, 255-267.

Ehret, D. L., and Ho, L. C. (1986). J. Exp. Bot. **37**, 1294-1302.

Eiseley, L. (1979). 'Darwin and the Mysterious Mr X'. J. M. Dent & Sons, London.

Emershad, R. L., and Ramming, D. W. (1984). Am. J. Bot. **71**, 873-877.

Engler, D. E., and Grogan, R. G. (1984). J. Hered. **75**, 426-430.

Eppendorfer, W. H. (1971). J. Sci. Food Agric. **22**, 501-505

Epstein, E. 'Mineral Nutrition of Plants: Principles and Perspectives'. J. Wiley & Sons, N. Y., London.

Esau, K. (1960). 'Anatomy of Seed Plants'. J. Wiley & Sons, N. Y., London.

Esau, K., and Thorsch, J. (1985). Am. J. Bot. **72**, 1641-1653.

Evanari, M. (1985). Annu. Rev. Plant Physiol. **36**, 1-25.

Evans, D. A., and Gamborg, O. L. (1982). Plant Cell Rep. **1**, 104-107.

Evans, D. A., Sharp, W. R., and Medina-Filho, H. P. (1984) Am. J. Bot. **71**, 759-774.

Evans, D. A., Sharp, W. R., and Bravo, J. E. (1987). Nestlé Research News 1986/87, 63-71.

Everson, R. G. (1970). Phytochem. **9**, 25-32.

Evert, R. F. (1984). In 'Contemporary Problems in Plant Anatomy' (R. A. White and W. C. Dickison, Eds), pp. 145-234. Academic Press, London.

Evstigneeva, Z. G., Pushkin, A. V., Akentyeva, N. P., and Kretovich, W. L. (1985). Physiol. Vég. **23**, 861-868.

Fader, G. M., and Koller, H. R. (1985). Plant Physiol. **79**, 663-666.

Fedoroff, N. V. (1983). In 'Mobile Genetic Elements' (J. A. Shapiro, Ed.), pp. 1-57. Academic Press, N.Y.

Fedoroff, N. V. (1984). Sci. Am. **250**, 64-74.

Felker, F. C. (1987). Am. J. Bot. **74**, 1912-1920.

Felker, F. C., and Shannon, J. C. (1980). Plant Physiol. **65**, 864-870.

Fellows, R. J., Egli, D. B., and Leggett, J. E. (1978). Plant Physiol. **62**, 812-814.

Ferguson, I. B. (1984). Plant, Cell Environ. **7**, 477-489.

Ferguson, I. B., and Bollard, E. G. (1976a). Ann. Bot. **40**, 1047-1055.

Ferguson, I. B., and Bollard, E. G. (1976b). Ann. Bot. **40**, 1057-1065.

Ferguson, J. D., Street, H. E., and David, S. B. (1958). Ann. Bot. **22**, 513-524.

Fisher, D. G. (1987). Aust. J. Plant Physiol. **14**, 147-155.

Fisher, D. G., and Frame, J. M. (1984). Planta **161**, 385-

393.

Fisher, D. G., and Gifford, R. M. (1986). Plant Physiol. **82**, 1024-1030.

Fisher, D. G., and Macnicol, P. K. (1986). Plant Physiol. **82**, 1019-1023.

Fisher, D. G., Larson, P. R., and Dickson, R. E. (1983). Bot. Gaz. **144**, 481-490.

Flinn, A. M., and Pate, J. S. (1968). Ann. Bot. **32**, 479-495.

Flinn, A. M., Atkins, C. A., and Pate, J. S. (1977). Plant Physiol. **60**, 412-418.

Flowers, T. J. (1972). J. Exp. Bot. **23**, 310-321.

Flügge, U. I., and Heldt, H. W. (1984). TIBS **9**, 530-533.

Fonda, M. L. (1972). Arch. Biochem. Biophys. **153**, 763-768.

Fraley, R. T., Rogers, S. G., and Horsch, R. B. (1987). CRC Critical Reviews in Plant Sci. **4**, 1-46.

Fraser, D. E., and Bidwell, R. G. S., (1974). Can. J. Bot. **52**, 2561-2570.

Fridovich, I. (1983). Annu. Rev. Pharmacol. Toxicol. **23**, 239-257.

Gamborg, O. L., and Shyluk, J. P. (1970). Plant Physiol. **45**, 598-600.

Gamborg, O. L., Shyluk, J. P., Brar, D. S., and Constabel, F. (1977). Plant Sci. Lett. **10**, 67-74.

Gengenbach, B. G., Green, C. E., and Donovan, C. M. (1977) Proc. Nat. Acad. Sci. U.S.A. **74**, 5113-5117.

Gengenbach, B. G., Connelly, J. A., Pring, D. R., and Conde, M. E. (1981). Theor. Appl. Genet. **59**, 161-167.

Gerhardt, R., Stitt, M., and Heldt, H. W. (1987). Plant Physiol. **83**, 399-407.

Giaquinta, R. T. (1976). Plant Physiol. **57**, 872-875.

Giaquinta, R. T. (1983). Annu. Rev. Plant Physiol. **34**, 347-387.

Gifford, R. M., and Bremner, P. M. (1981a). Aust. J. Plant Physiol. **8**, 619-629.

Gifford, R. M., and Bremner, P. M. (1981b). Aust. J. Plant Physiol. **8**, 631-640.

Gifford, R. M., and Thorne, J. H. (1985). Plant Physiol. **77**, 863-868.

Gifford, R. M., and Thorne, J. H. (1986). Plant Physiol. **80**, 464-469.

Gimmler, H., Schäfer, G., Kraminer, H., and Heber, U. (1974). Planta **120**, 47-64.

Givan, C. V. (1975). Planta **122**, 281-291.

Goodman, M., Benson, A. A., and Calvin, M. (1955). J. Am. Chem. Soc. **77**, 4257.

Gordon, K. H. J., Peoples, M. B., and Murray, D. R. (1978) New Phytol. **81**, 35-42.

Govindjee, Wyndrzynski, T., and Marks, S. B. (1978). In 'Photosynthetic Oxygen Evolution' (H. Metzner, Ed.), pp. 321-344. Academic Press, New York.

Goyal, A., and Gupta, R. K. (1985). J. Nuclear Agric. Biol. **14**, 4-7.

Grandbastien, M.-A. (1987). Nestlé Research News 1986/87, 31-45.

Grobbelaar, N., and Steward, F. C. (1958). Nature **182**, 1358-1359.

Groussol, J., Delrot, S., Caruhel, P., and Bonnemain, J. -L. (1986). Physiol. Vég. **24**, 123-133.

Guardiola, J. L., and Sutcliffe, J. F. (1972). J. Exp. Bot. **23**, 322-337.

Gunning, B. E. S., and Robards, A. W. (1976). In 'Transport and Transfer Processes in Plants' (I. F. Wardlaw and J. B. Passioura, Eds), pp. 15-41. Academic Press, New York.

Gunning, B. E. S., Pate, J. S., and Green, L. W. (1970). Protoplasma **71**, 147-171.

Gustafson, F. G., and Darken, M. (1937). Am. J. Bot. **24**, 615-621.

Haccius, B., and Lakshmanan, K. K. (1969). Österr. Bot. Z. **116**, 145-158.

Hall, D. O., Rao, K. K., and Cammack, R. (1975). Sci. Prog. Oxford **62**, 285-317.

Hall, S. M., and Milburn, J. A. (1973). Planta **109**, 1-10.

Halliwell, B. (1974). New Phytol. **73**, 1075-1086.

Halperin, W., and Wetherell, D. F. (1965). Nature **205**, 519-520.

Hanson, A. D., and Tully, R. E. (1979). Plant Physiol. **64**, 467-471.

Harborne, J. B. (1982). 'Introduction to Ecological Biochemistry'. 2nd edition. Academic Press, London.

Hardham, A. R. (1976). Aust. J. Bot. **24**, 711-721.

Harel, E., Lea, P. J., and Miflin, B. J. (1979). Planta

134, 195-200.

Harper, R. J., Gilkes, R. J., and Robson, A. D. (1982). Aust. J. Agric. Res. **33**, 565-571.

Harris, H. C. (1949). Plant Physiol. **24**, 150-161.

Harris, W. M. (1983). Am. J. Bot. **70**, 1528-1535.

Hartmann, T., and Ehmke, A. (1980). Planta **149**, 207-208.

Hartt, C. E. (1965). Plant Physiol. **40**, 718-724.

Hartt, C. E., and Kortschak, H. P. (1964). Plant Physiol. **39**, 460-474.

Hartt, C. E., and Kortschak, H. P. (1967). Plant Physiol. **42**, 89-94.

Harvey, D. M. (1973). Ann. Bot. **37**, 787-794.

Harvey, D. M., Hedley, C. L., and Keely, R. (1976). Ann. Bot. **40**, 993-1001.

Hatch, M. D., and Slack, C. R. (1966). Biochem. J. **101**, 103-111.

Hatch, M. D., Kagawa, T., and Craig, S. (1975). Aust. J. Plant Physiol. **2**, 111-128.

Hattersley, P. W., Watson, L., and Osmond, C. B. (1977). Aust. J. Plant Physiol. **4**, 523-539.

Hawker, J. S. (1965). Aust. J. Biol. Sci. **18**, 959-969.

Hawker, J. S. (1967). Biochem. J. **105**, 943-946.

Hawker, J. S. (1969). Phytochem. **8**, 9-17.

Hawker, J. S. (1971). Phytochem. **10**, 2313-2322.

Hawker, J. S. (1985). In 'Biochemistry of Storage Carbohydrates in Green Plants' (P. M. Dey and R. A. Dixon, Eds), pp. 1-51. Academic Press, London.

Hawker, J. S., Ruffner, H. P., and Walker, R. R. (1976). Am. J. Enol. Vitic. **27**, 125-129.

Heber, U., Kirk, M. R., Gimmler, H., and Schaeffer, F. Planta **120**, 31-46.

Hedley, C. L., Harvey, D. M., and Keely, R. (1975). Nature **258**, 352-353.

Heinz, D. J., and Mee, G. W. P. (1971). Am. J. Bot. **58**, 257-262.

Heinz, D. J., Krishnamurthi, M., Nickell, L. G., and Maretzki, A. (1977). In 'Applied and Fundamental Aspects of Plant Cell, Tissue and Organ Culture' (J. Reinert and Y. P. S. Bajaj, Eds), pp. 3-17. Springer-Verlag, Berlin.

Hein, M. B., Brenner, M. L., and Brun, W. A. (1984). Plant Physiol. **76**, 951-954.

Hein, M. B., Brenner, M. L., and Brun, W. A. (1986). Plant Physiol. **82**, 454-456.

Heldt, H. W. (1979). In 'Encyclopaedia of Plant Physiology New Series Vol. **6** (M. Gibbs and E. Latzko, Eds), pp. 202-207. Springer-Verlag, Berlin.

Henderson, J. H. M., Phills, B. R., and Whatley, B. T. (1984). In 'Handbook of Plant Cell Culture Vol. 2 Crop Species' (W. R. Sharp, D. A. Evans, P. V. Ammirato and Y. Yamada, Eds), pp. 302-326. Macmillan Pub. Co., N. Y.

Hepler, P. K., and Wayne, R. O. (1985). Annu. Rev. Plant Physiol. **36**, 397-439.

Herridge, D. F. (1982a). Plant Physiol. **70**, 1-5.

Herridge, D. F. (1982b). Plant Physiol. **70**, 7-11.

Herridge, D. F., and Pate, J. S. (1977). Plant Physiol. **60**, 759-764.

Herridge, D. F., Atkins, C. A., Pate, J. S., and Rainbird, R. M. (1978). Plant Physiol. **62**, 495-498.

Hewitt, E. J. (1975). Annu. Rev. Plant Physiol. **26**, 73-100.

Ho, L. C., and Mortimer, D. C. (1971). Can. J. Bot. **49**, 1769-1775.

Ho, L. C., and Peel, A. J. (1969). Physiol. Plant. **22**, 379-385.

Hoad, G. V., and Peel, A. J. (1965). J. Exp. Bot. **16**, 433-451.

Hocking, P. J. (1980). Ann. Bot. **45**, 633-643.

Hocking, P. J. (1982). Ann. Bot. **49**, 51-62.

Hocking, P. J. (1983a). Aust. J. Agric. Res. **34**, 155-166.

Hocking, P. J. (1983b). New Phytol. **93**, 511-529.

Hocking, P. J., and Pate, J. S. (1977). Ann. Bot. **41**, 1259-1278.

Hocking, P. J., Pate, J. S., Wee, S. C., and McComb, A. J. (1977). Ann. Bot. **41**, 677-688.

Hocking, P. J., Pate, J. S., Atkins, C. A., and Sharkey, P. J. (1978). Ann. Bot. **42**, 1277-1290.

Holowach, L. P., Thompson, J. F., and Madison, J. T. (1984a). Plant Physiol. **74**, 576-583.

Holowach, L. P., Thompson, J. F., and Madison, J. T. (1984b). Plant Physiol. **74**, 584-589.

Hooda, R. S., Rao, A. S., Luthra, Y. P., Sheoran, I. S., and Singh, R. (1986). J. Exp. Bot. **37**, 1492-1502.

Howe, H. F. (1986). In 'Seed Dispersal' (D. R. Murray,

Ed.), pp. 123-189. Academic Press, Sydney.

Howitz, K. T., and McCarty, R. E. (1986). Plant Physiol. **80**, 390-395.

Hsu, F. C., Bennett, A. B., and Spanswick, R. M. (1984). Plant Physiol. **75**, 181-186.

Huber, S. C. (1986). Annu. Rev. Plant Physiol. **37**, 233-246.

Huber, S. C., and Akazawa, T. (1986). Plant Physiol. **81**, 1008-1013.

Huber, S. C., and Edwards, G. E. (1975). Plant Physiol. **56**, 324-331.

Huber, S. C., Hall, T. C., and Edwards, G. E. (1976). Plant Physiol. **57**, 730-733.

Hughes, M. N. (1981). 'The Inorganic Chemistry of Biological Processes'. 2nd edition. John Wiley & Sons, Chichester.

Ireland, R. J., and Joy, K. W. (1981). Planta **151**, 289-292.

Ireland, R. J., and Joy, K. W. (1983). Plant Physiol. **72**, 1127-1129.

Israel, D. W., and Jackson, W. A. (1978). In 'Mineral Nutrition of Legumes in Tropical and Subtropical Soils' (C. S. Andrew and E. J. Kamprath, Eds), pp. 113-129. CSIRO, Melbourne Australia.

Janardhanan, K., and Lakshmanan, K. K. (1982a). Current Science **51**, 756-757.

Janardhanan, K., and Lakshmanan, K. K. (1982b). Ind. J. Forestry **5**, 255-259.

Jenner, C. F. (1974). Royal Soc. N.Z. Bull. **12**, 901-908.

Jenner, C. F. (1982). Aust. J. Plant Physiol. **9**, 69-82.

Johnson, A. D., and Simons, J. G. (1979). J. Plant Nutr. **1**, 123-149.

Johnston, M., Grof, C. P. L., and Brownell, P. F. (1984). Aust. J. Plant Physiol. **11**, 137-141.

Johri, B. M. (1984). 'Embryology of Angiosperms'. Springer-Verlag, Berlin.

Jones, R. K., and Clay, H. J. (1976). 'Foliar Symptoms of Mineral Disorders in Townsville Stylo (*Stylosanthes humilis*)'. Division of Tropical Crops and Pastures Technical Paper No. **19**. CSIRO, Australia.

Kallarackal, J., and Milburn, J. A. (1983). Aust. J. Plant Physiol. **10**, 561-568.

Kallarackal, J., and Milburn, J. A. (1984). Aust. J. Plant Physiol. **11**, 483-490.

Kallarackal, J., and Milburn, J. A. (1985a). Plant, Cell

Environ. **8**, 239-245.

Kallarackal, J., and Milburn, J. A. (1985b). Ann. Bot. **56**, 211-218.

Kawabe, S., Fukumorita, T., and Chino, M. (1980). Plant Cell Physiol. **21**, 1319-1327.

Kearney, P. C., and Tolbert, N. E. (1962). Arch. Biochem. Biophys. **98**, 164-171.

Kelly, G. J., and Turner, J. F. (1969). Biochem. J. **115**, 481-487.

Kennedy, I. R. (1986). 'Acid Soil and Acid Rain - The Impact on the Environment of Nitrogen and Sulphur Cycling'. Research Studies Press, John Wiley & Sons, Chichester, U. K.

Kennedy, J. S., and Mittler, T. E. (1953). Nature **171**, 528.

Kerr, P. S., Blevins, D. G., Rapp, B. J., and Randall, D. D. (1983). Physiol. Plant. **57**, 339-345.

Keys, A. J. (1980). In 'The Biochemistry of Plants Vol. 5 Amino Acids and Derivatives' (B. J. Miflin, Ed.), pp. 359-374. Academic Press, N. Y.

King, R. W., and Zeevaart, J. A. D. (1974). Plant Physiol. **53**, 96-103.

Kipps, A., and Boulter, D. (1973). New Phytol. **72**, 1293-1297.

Kipps, A., and Boulter, D. (1974). New Phytol. **73**, 675-684.

Kisaki, T., and Tolbert, N. E. (1969). Plant Physiol. **44**, 242-250.

Knypl, J. S., and Ferguson, A. R. (1975). Z. Pflanzenphysiol. **74**, 434-439.

Kortschak, H. P., Hartt, C. E., and Burr, G. O. (1965). Plant Physiol. **40**, 209-213.

Krikorian, A. D., and Cronauer, S. S. (1984). In 'Handbook of Plant Cell Culture Vol. 2 Crop Species (W. R. Sharp, D. A. Evans, P. V. Ammirato and Y. Yamada, Eds), pp. 327-348. Macmillan Pub. Co., N. Y.

Krishna, T. G., and Murray, D. R. (1988). J. Plant Physiol (in press).

Krishnamurthi, M., and Tlaskal, J. (1974). Proc. Int. Soc. Sugar Cane Technol. **15**, 130-137.

Kumar, K. V., Janardhanan, K., and Lakshmanan, K. K. (1983). Ind. Bot. Contactor **1**, 131-132.

Kumarasinghe, K. S., Keys, A. J., and Whittingham, C. P. (1977). J. Exp. Bot. **28**, 1247-1257.

Kurvits, A., and Kirkby, E. A. (1980). Z. Pflanzenernährung und Bodenkunde **143**, 140-149.

Laing, W. A., and Forde, B. J. (1971). Planta **98**, 221-231.

Laing, W. A., Ogren, W. L., and Hageman, R. H. (1974). Plant Physiol. **54**, 678-685.

Lakshmanan, K. K. (1969). Ind. Biol. **1**, 16-17.

Lakshmanan, K. K. (1975). 'Tissue Culture'. Tamil Nadu Textbook Society, Madras.

Lakshmanan, K. K., and Janardhanan, K. (1983). S. B. Cl. Newsletter **2**, 55-61.

Larkin, P. J. (1985). In 'Cereal Tissue and Cell Culture' (S. W. J. Bright and M. G. K. Jones, Eds), pp. 274-296. M. Nijhoff/W. Junk Pub., Dordrecht.

Larkin, P. J. (1987). Iowa State J. Res. **61**, 393-434.

Larkin, P. J., and Scowcroft, W. R. (1981a). Theor. Appl. Genet. **60**, 197-214.

Larkin, P. J., and Scowcroft, W. R. (1981b). Plant Physiol. **67**, 408-414.

Larkin, P. J., and Scowcroft, W. R. (1983a). Plant Cell Tissue Organ Culture **2**, 111-121.

Larkin, P. J., and Scowcroft, W. R. (1983b). In 'Genetic Engineering of Plants' (T. Kosuge, C. P. Meredith and A. Hollaender, Eds), pp. 289-314. Plenum Pub. Corp.

Larkin, P. J., Brettell, R. I. S., Ryan, S. A., and Scowcroft, W. R. (1983). 'Protoplasts 1983'. pp. 51-56. Birhäuser Verlag, Basel.

Larkin, P. J., Ryan, S. A., Brettell, R. I. S., and Scowcroft, W. R. (1984). Theor. Appl. Genet. **67**, 443-455.

Lastra, O., Gomez, J., Lopez-Gorge, J., and del Río, L. A. (1982). Physiol. Plant. **55**, 209-213.

Lawlor, D. W. (1978). In 'Stress Physiology in Crop Plants' (H. Mansell and R. C. Staples, Eds), pp. Wiley Interscience, New York.

Lea, P. J., and Miflin, B. J. (1974). Nature **215**, 614-616.

Lea, P. J., Hughes, J. S., and Miflin, B. J. (1979). J. Exp. Bot. **30**, 529-537.

Leopold, A. C., and Kriedemann, P. E. (1975). 'Plant Growth and Development'. 2nd edition. McGraw-Hill, N.Y.

Lersten, N. R. (1983). New Phytol. **93**, 633-637.

Lewis, O. A. M. (1975). J. Exp. Bot. **26**, 361-366.

Lewis, O. A. M., and Pate, J. S. (1973). J. Exp. Bot. **24**, 596-606.

Lichtner, F. T., and Spanswick, R. M. (1981). Plant Physiol. **68**, 693-698.

Linck, A. J., and Sudia, T. W. (1962). Experientia **18**, 69-70.

Linck, A. J., and Swanson, C. A. (1960). Plant and Soil **12**, 57-68.

Lingle, S. E., and Dunlap, J. R. (1987). Plant Physiol. **84**, 386-389.

Linsmaier, E. M., and Skoog, F. (1965). Physiol. Plant. **18**, 100-127.

Liu, M.-C. (1981). In 'Plant Tissue Culture - Methods and Applications in Agriculture' (T. A. Thorpe, Ed.), pp. 299-323. Academic Press, New York.

Loneragan, J. F., Snowball, K., and Robson, A. D. (1976). In 'Transport and Transfer Processes in Plants' (I. F. Wardlaw and J. B. Passioura, Eds), pp. 463-469. Academic Press, New York.

Lorenzi, R., Bennici, A., Cionini, P. G., Alpi, A., and D'Amato, F. (1978). Planta **143**, 59-62.

Lorimer, G. (1981). Annu. Rev. Plant Physiol. **32**, 349-383.

Lorimer, G., Andrews, T. J., and Tolbert, N. E. (1973). Biochem. **12**, 18-23.

Lott, J. N. A. (1984). In 'Seed Physiology Vol. 1 Development' (D. R. Murray, Ed.), pp. 139-166. Academic Press, Sydney.

Lott, J. N. A., and Volmer, C. M. (1979). Can. J. Bot. **57**, 987-992.

Lott, J. N. A., Goodchild, D. J., and Craig, S. (1984). Aust. J. Plant Physiol. **11**, 459-469.

Lovatt, C. J. (1983). Plant Physiol. **73**, 766-772.

Lovell, P. H., and Lovell, P. J. (1970). Physiol. Plant. **23**, 316-322.

Lovell, P. H., Oo, H. T., and Sagar, G. R. (1972). J. Exp. Bot. **23**, 255-266.

Lowry, O. H., Rosebrough, N. J., Farr, A. L., and Randall, R. J. (1951). J. Biol Chem. **193**, 265-275.

Ludwig, L. J., and Canvin, D. T. (1971). Plant Physiol. **48**, 712-719.

Macklon, A. E. S., and Sim, A. (1980). Physiol. Plant. **49**, 449-454.

Macnicol, P. K., and Bergmann, L. (1984). Plant Sci. Lett. **36**, 219-223.

Macnicol, P. K., and Randall, P. J. (1987). Plant Physiol.

83, 354-359.

Marbach, I., and Mayer, A. M. (1974). Plant Physiol. **54**, 817-820.

Marbach, I., and Mayer, A. M. (1975). Plant Physiol. **56**, 93-96.

Marcelle, R., and Bodson, M. (1979). J. Plant Nutr. **1**, 207-217.

Marshall, C., and Sagar, G. R. (1978). In 'Plant Structure Function and Adaptation' (M. A. Hall, Ed.), pp. 254-293. Macmillan, London.

Martin, F. A., Ozbun, J. L., and Wallace, D. H. (1972). Plant Physiol. **49**, 764-768.

Martin, P. (1971). Z. Pflanzenphysiol. **64**, 206-222.

Mason, T. G., and Maskell, E. J. (1931). Ann. Bot. **XLV**, 126-173.

Matheson, N. K. (1984). In 'Seed Physiology Vol. 1 Development' (D. R. Murray, Ed.), pp. 167-208. Academic Press, Sydney.

Mathews, V. H., Bhatia, C. R., Mitra, R., Krishna, T. G., and Rao, P. S. (1985). Plant Sci. **39**, 49-54.

Mauney, J. R. (1961). Bot. Gaz. **122**, 205-209.

Mauney, J. R., Chappell, J., and Ward, B. J. (1967). Bot. Gaz. **128**, 198-200.

Mazzolini, A. P., Pallaghy, C. K., and Legge, G. J. F. (1985). New Phytol. **100**, 483-509.

McClintock, B. (1950). Proc. Nat. Acad. Sci. U.S.A. **36**, 344-355.

McClure, P. R., and Israel, D. W. (1979). Plant Physiol. **64**, 411-416.

McCoy, T. J., Phillips, R. L., and Rines, H. W. (1982). Can. J. Genet. Cytol. **24**, 37-50.

McGee, C. M., and Murray, D. R. (1985). J. Plant Physiol. **120**, 1-8.

McGee, C. M., and Murray, D. R. (1986). Ann. Bot. **57**, 179-190.

McKee, H. S. (1962). 'Nitrogen Metabolism in Plants'. Oxford University Press.

McNeil, D. L. (1980). Ann. Bot. **45**, 329-338.

McNeil, D. L., Atkins, C. A., and Pate, J. S. (1979). Plant Physiol. **63**, 1076-1081.

Melcher, I. M. (1985). New Phytol. **100**, 157-162.

Miflin, B. J., Marker, A. F. H., and Whittingham, C. P. (1966). Biochim. Biophys. Acta **120**, 266-273.

Milburn, J. A. (1970). Planta **95**, 272-276.

Milburn, J. A. (1971). Planta **100**, 143-155.

Milburn, J. A. (1972). Pesticide Sci. **3**, 653-665.

Milburn, J. A. (1974). Planta **117**, 303-319.

Milburn, J. A. (1979). 'Water Flow in Plants'. Longman, London, New York.

Milburn, J. A., and Kallarackal, J. (1984). New Phytol. **96**, 383-395.

Millikan, C. R., and Hanger, B. C. (1965). Aust. J. Biol. Sci. **18**, 211-226.

Minchin, P. E. H., and McNaughton, G. S. (1986). J. Exp. Bot. **37**, 1151-1163.

Minchin, P. E. H., and Thorpe, M. R. (1984). J. Exp. Bot. **35**, 538-550.

Misra, S., and Oaks, A. (1985). Plant Physiol. **77**, 520-523.

Mitchell, C. A., and Stocking, C. R. (1975). Plant Physiol. **55**, 59-63.

Mittler, T. E. (1957). J. Exp. Biol. **34**, 334-341.

Mix, G. P., and Marschner, H. (1976). Z. Pflanzenphysiol. **80**, 354-366.

Molin, W. T., and Fites, R. C. (1980). Plant Physiol. **66**, 308-312.

Münch, E. (1930). 'Die Stoffbewegung in der Pflanze'. Fischer, Jena.

Munns, R., and Passioura, J. B. (1984). Aust. J. Plant Physiol. **11**, 497-507.

Murashige, T. (1974). Annu. Rev. Plant Physiol. **25**, 135-166.

Murray, D. R. (1969). 'Photoassimilation of Glycolate and Related Compounds by *Euglena gracilis*'. Ph.D. Thesis, The University of Sydney.

Murray, D. R. (1979a). Plant, Cell Environ. **2**, 221-226.

Murray, D. R. (1979b). Plant Physiol. **64**, 763-769.

Murray, D. R. (1980a). Ann. Bot. **45**, 273-281.

Murray, D. R. (1980b). Ann. Bot. **46**, 499-504.

Murray, D. R. (1983a). Z. Pflanzenphysiol. **110**, 7-15.

Murray, D. R. (1983b). New Phytol. **93**, 33-41.

Murray, D. R. (1984a). In 'Seed Physiology Vol. 1 Development' (D. R. Murray, Ed.), pp. 1-40. Academic Press, Sydney.

Murray, D. R. (1984b). In 'Seed Physiology Vol. 1 Development' (D. R. Murray, Ed.), pp. 83-137. Academic Press, Sydney.

Murray, D. R. (1985). New Phytol. **101**, 377-385.

Murray, D. R. (1986a). New Phytol. **104**, 395-406.

Murray, D. R. (1986b). 'Guidelines for the Quantitative Estimation of Protein in Plant Tissue Extracts'. Dept. of Biology Technical Report No. **4**, University of Wollongong.

Murray, D. R. (1986c). In 'Seed Dispersal' (D. R. Murray, Ed.), pp. 49-85. Academic Press, Sydney.

Murray, D. R. (1987). Am. J. Bot. **74**, 1122-1137.

Murray, D. R., and Ayre, D. J. (1987). J. Plant Physiol. **127**, 193-201.

Murray, D. R., and Collier, M. D. (1977). Aust. J. Plant Physiol. **4**, 843-848.

Murray, D. R., and Cordova-Edwards, M. (1984a). New Phytol. **97**, 243-252.

Murray, D. R., and Cordova-Edwards, M. (1984b). New Phytol. **97**, 253-260.

Murray, D. R., and Kennedy, I. R. (1980). Plant Physiol. **66**, 782-786.

Murray, D. R., Giovanelli, J., and Smillie, R. M. (1971). Aust. J. Biol. Sci. **24**, 23-33.

Murray, D. R., Wara-Aswapati, O., Ireland, H. M. M., and Bradbeer, J. W. (1973). J. Exp. Bot. **24**, 175-184.

Nable, R. O., and Loneragan, J. F. (1984). Aust. J. Plant Physiol. **11**, 101-111.

Nagl, W. (1976a). Ber. Deut. Bot. Ges. **89**, 301-311.

Nagl, W. (1976b). Biochem. Physiol. Pflanzen **170**, 255-260.

Nakamura, Y., and Tolbert, N. E. (1983). J. Biol. Chem. **258**, 7631-7638.

Nason, A., and McElroy, W. D. (1963). In 'Plant Physiology Vol. III. Inorganic Nutrition of Plants' (F. C. Steward, Ed.), pp. 451-536. Academic Press, New York.

Neeman, M., Aviv, D., Degani, H., and Galun, E. (1985). Plant Physiol. **77**, 374-378.

Nelson, C. D. (1962). Can. J. Bot. **40**, 757-770.

Nelson, C. D., Clauss, H., Mortimer, D. C., and Gorham, P. R. (1961). Plant Physiol. **36**, 581-588.

Nelson-Schreiber, B. M., and Schweitzer, L. E. (1986). Plant Physiol. **80**, 454-458.

Neumann, P. M., and Noodén, L. D. (1984). Physiol. Plant. **60**, 166-170.

Newcomb, W. (1973). Can. J. Bot. **51**, 879-890.

Neyra, C. A., and Hageman, R. H. (1978). Plant Physiol. **62**, 618-621.

Nguyen, J. (1986). Physiol. Vég. **24**, 263-281.

Nicholas, D. J. D. (1961). Annu. Rev. Plant Physiol. **12**, 63-90.

Nicholas, D. J. D., and Nason, A. (1954). J. Biol. Chem. **207**, 353-360.

Nitsos, R. E., and Evans, H. J. (1969). Plant Physiol. **44**, 1260-1266.

Noggle, G. R., and Fritz, G. J. (1976). 'Introductory Plant Physiology'.Prentice-Hall Inc., New Jersey.

Noodén, L. D. (1984). Physiol. Plant. **62**, 273-284.

Nutman, P. S. (1939). Ann. Bot. **3**, 731-757.

Nutman, P. S. (1976). 'Symbiotic Nitrogen Fixation in Plants'. I.B.P. Vol. **7**. Cambridge University Press.

O'Dowd, D. J., and Gill, A. M. (1986). In 'Seed Dispersal' (D. R. Murray, Ed.), pp. 87-121. Academic Press, Sydney.

Offler, C. E., and Patrick, J. W. (1984). Aust. J. Plant Physiol. **11**, 79-99.

Offler, C. E., and Patrick, J. W. (1986). In 'Phloem Transport'. Plant Biology Vol. **1**. (J. Cronshaw, W. J. Lucas and R. T. Giaquinta, Eds), pp. 295-306. A. R. Liss, N.Y.

Ogren, W. L. (1984). Annu. Rev. Plant Physiol. **35**, 415-442.

O'Neal, T. D., and Joy, K. W. (1974). Plant Physiol. **54**, 773-779.

Oono, K. (1981). In 'Plant Tissue Culture - Methods and Application in Agriculture' (T. A. Thorpe, Ed.), pp. 273-298. Academic Press, London.

Oparka, K. J., and Gates, P. (1981a). Planta **151**, 561-573.

Oparka, K. J., and Gates, P. (1981b). Planta **152**, 388-396.

Oparka, K. J., and Gates, P. (1982). Protoplasma **113**, 33-43.

Owens, L. D. (1979a). Plant Sci. Lett. **16**, 225-230.

Owens, L. D. (1979b). Plant Physiol. **63**, 683-686.

Owens, L. D. (1981). Plant Physiol. **67**, 1166-1168.

Pate, J. S. (1966). Ann. Bot. **30**, 93-109.

Pate, J. S. (1973). Soil Biol. Biochem. **5**, 109-119.

Pate, J. S. (1976a). In 'Transport and Transfer Processes in Plants' (I. F. Wardlaw and J. B. Passioura, Eds), pp. 253-281. Academic Press, New York.

Pate, J. S. (1976b). In 'Transport and Transfer Processes in Plants' (I. F. Wardlaw and J. B. Passioura, Eds), pp. 447-462. Academic Press, New York.

Pate, J. S. (1977). In 'The Physiology of the Garden Pea' (J. F. Sutcliffe and J. S. Pate, Eds), pp. 349-383. Academic Press, London.

Pate, J. S. (1980). Annu. Rev. Plant Physiol. **31**, 313-340.

Pate, J. S. (1984). In 'Seed Physiology Vol. 1. Development' (D. R. Murray, Ed.), pp. 41-82.

Pate, J. S., and Herridge, D. F. (1978). J. Exp. Bot. **29**, 401-412.

Pate, J. S., and Hocking, P. J. (1978). Ann. Bot. **42**, 911-921.

Pate, J. S., and Wallace, W. (1964). Ann. Bot. **28**, 80-99.

Pate, J. S., Gunning, B. E. S., and Milliken, F. F. (1970) Protoplasma **71**, 313-334.

Pate, J. S., Sharkey, P. J., and Lewis, O. A. M. (1974). Planta **120**, 229-243.

Pate, J. S., Sharkey, P. J., and Lewis, O. A. M. (1975). Planta **122**, 11-26.

Pate, J. S., Sharkey, P. J., and Atkins, C. A. (1977). Plant Physiol. **59**, 506-510.

Pate, J. S., Kuo, J., and Hocking, P. J. (1978). Aust. J. Plant Physiol. **5**, 321-336.

Pate, J. S., Layzell, D. B., and Atkins, C. A. (1979a). Plant Physiol. **64**, 1083-1088.

Pate, J. S., Layzell, D. B., and McNeil, D. M. (1979b). Plant Physiol. **63**, 730-738.

Pate, J. S., Peoples, M. B., and Atkins, C. A. (1984). Plant Physiol. **74**, 499-505.

Pate, J. S., Peoples, M. B., Van Bel, A. J. E., Kuo, J., and Atkins, C. A. (1985). Plant Physiol. **77**, 148-156.

Patrick, J. W. (1981). Aust. J. Plant Physiol. **8**, 221-235.

Patrick, J. W. (1983). Z. Pflanzenphysiol. **111**, 9-18.

Patrick, J. W. (1984). J. Plant Physiol. **115**, 297-310.

Patrick, J. W., and McDonald, R. (1980). Aust. J. Plant Physiol. **7**, 671-684.

Pearson, C. J. (1974). Planta **119**, 59-70.

Peel, A. J. (1963). J. Exp. Bot. **14**, 438-447.

Pelletier, G. (1987). Nestlé Research News 1986/87, 55-61.

Peoples, M. B., Beilharz, V. C., Waters, S. P., Simpson, R. J., and Dalling, M. J. (1980). Planta **149**, 241-251.

Peoples, M. B., Pate, J. S., and Atkins, C. A. (1983). J. Exp. Bot. **34**, 563-578.

Peoples, M. B., Atkins, C. A., Pate, J. S., and Murray, D. R. (1985a). Plant Physiol. **77**, 382-388.

Peoples, M. B., Pate, J. S., Atkins, C. A., and Murray, D. R. (1985b). Plant Physiol. **77**, 142-147.

Peoples, M. B., Pate, J. S., Atkins, C. A., and Bergersen, F. J. (1986). Plant Physiol. **82**, 946-951.

Peoples, M. B., Sudin, M. N., and Herridge, D. F. (1987). J. Exp. Bot. **38**, 567-579.

Peters, J. A. (1959). 'Classic Papers in Genetics'. Prentice-Hall Inc., Englewood Cliffs, New Jersey.

Pfeiffer, J. W., and Bingham, E. T. (1984). Theor. Appl. Genet. **67**, 263-266.

Pissarek, H.-P. (1980). Z. Pflanzenernährung und Bodenkunde **143**, 150-160.

Plant, A. R., and Moore, K. G. (1983). New Phytol. **93**, 359-368.

Pless, T., Böttger, M., Hedden, P., and Graebe, J. (1984). Plant Physiol. **74**, 320-323.

Plummer, D. T., Murray, D. R., and Linnane, A. W. (1974). Biochem. Soc. Trans. **2**, 221-223.

Polacco, J. C., and Sparks, R. B. Jr. (1982). Plant Physiol. **70**, 189-194.

Pollak, P. E., and Mogensen, H. L. (1986). Am. J. Bot. **73**, 48-59.

Porter, G. A., Knievel, D. P., and Shannon, J. C. (1985). Plant Physiol. **77**, 524-531.

Portis, A. P. Jr., Salvucci, M. E., and Ogren, W. L. (1986). Plant Physiol. **82**, 967-971.

Portnoi, L., and Horovitz, A. (1977). Ann. Bot. **41**, 21-27.

Power, C. J. (1987). Am. J. Bot. **74**, 497-503.

Powles, S. B. (1984). Annu. Rev. Plant Physiol. **35**, 15-44.

Powles, S. B., Osmond, C. B., and Thorne, S. W. (1979). Plant Physiol. **64**, 982-988.

Preiss, J. (1982). Annu. Rev. Plant Physiol. **33**, 431-454.

Preiss, J. (1984). TIBS **9**, 24-27.

Price, H. J., and Smith, R. H. (1979). Planta **145**, 305-307.

Przybyllok, T., and Nagl, W. (1977). Z. Pflanzenphysiol. **84**, 463-465.

Qi-gui, C., Zhongchen, G., Ying-qian, Q., Rong-xi, J., and Yun-luo, Z. (1987). Acta Bot. Sinica **29**, 453-458.

Raacke, I. D. (1957). Biochem. J. **66**, 110-113.

Radin, J. W. (1978). Plant Sci. Lett. **13**, 21-25.

Raghavan, V. (1977). In CRC Handbook Series in Nutrition and Food (M. Rechcigl Jr., Ed.), Section G: Diets, Culture Media and Food Supplements Vol. **IV**, pp. 361-413. CRC Press, Cleveland.

Rainbird, R. M., Thorne, J. H., and Hardy, R. W. F. (1984) Plant Physiol. **74**, 329-334.

Raju, A. R., Janardhanan, K., and Lakshmanan, K. K. (1983) Ind. J. Exp. Biol. **21**, 633-634.

Ramshaw, J. A. M., Brown, R. H., Scawen, M. D., and Boulter, D. (1973). Biochim. Biophys. Acta **459**, 418-441.

Randall, P. J., and Bouma, D. (1973). Plant Physiol. **52**, 229-232.

Randall, P. J., Thomson, J. A., and Schroeder, H. E. (1979). Aust. J. Plant Physiol. **6**, 11-24.

Ray, T. B. (1984). Plant Physiol. **75**, 827-831.

Reed, J. E., and Chollet, R. (1985). Planta **165**, 439-445.

Reisch, B., and Bingham, E. T. (1981). Crop Sci. **21**, 783-788.

Richardson, K. E., and Tolbert, N. E. (1961a). J. Biol. Chem. **236**, 1280-1284.

Richardson, K. E., and Tolbert, N. E. (1961b). J. Biol. Chem. **236**, 1285-1290.

Rijven, A. H. G. C. (1952). Acta Bot. Neerl. **1**, 157-200.

Rijven, A. H. G. C. (1956). Aust. J. Biol. Sci. **9**, 511-527.

Rijven, A. H. G. C., and Gifford, R. M. (1983). Plant Cell Environ. **6**, 417-425.

Rinne, R. W., and Langston, R. G. (1960). Plant Physiol. **35**, 216-219.

Robinson, S. P., and Downton, W. J. S. (1985). Aust. J. Plant Physiol. **12**, 471-479.

Rocha, V., and Ting, I. P. (1971). Arch. Biochem. Biophys. **147**, 114-122.

Rosenthal, G. A. (1982). 'Plant Non-Protein Amino and

Imino Acids'. Academic Press, New York.

Rosenthal, G. A., and Rhodes, D. (1984). Plant Physiol. **76**, 541-544.

Ross, G. S., Minchin, P. E. H., and McWha, J. A. (1988). J. Plant Physiol.

Rufty, T. W. Jr., Raper, C. D. Jr., and Jackson, W. A. (1983). Bot. Gaz. **144**, 466-470.

Rufty, T. W. Jr., Thomas, J. F., Remmler, J. L., Campbell, W. H., and Volk, R. J. (1986). Plant Physiol. **82**, 675-680.

Russel, C. R., and Morris, D. A. (1982). Ann. Bot. **49**, 89-98.

Ryan, S. A., and Scowcroft, W. R. (1987). Theor. Appl. Genet. **73**, 459-464.

Ryczkowski, M. (1960a). Planta **55**, 343-356.

Ryczkowski, M. (1960b). Planta **55**, 357-364.

Ryczkowski, M. (1962). Acta Soc. Bot. Polon. **31**, 53-65.

Ryczkowski, M. (1964a). Acta Soc. Bot. Polon. **33**, 397-406.

Ryczkowski, M. (1964b). Bull. Acad. Polon. Sci. Ser. Sci. Biol. **12**, 269-275.

Ryczkowski, M. (1967). Acta Soc. Bot. Polon. **36**, 627-638.

Ryczkowski, M. (1969). Z. Pflanzenphysiol. **61**, 422-429.

Sacher, J. A. (1966). Plant Physiol. **41**, 181-189.

Sachs, J. (1875). 'Textbook of Botany - Morphological and Physiological' (A. W. Bennett and W. T. T. Dyer, translators). The University of Oxford (Clarendon Press), MacMillan & Co., London.

Sacristan, M. D. (1982). Theor. Appl. Genet. **61**, 193-200.

Sakri, F. A. K., and Shannon, J. C. (1975). Plant Physiol. **55**, 581-589.

Salin, M. L., and Bridges, S. M. (1982). Plant Physiol. **69**, 161-165.

Salisbury, F. B., and Ross, C. W. (1978). 'Plant Physiology'. 2nd edition. Wadsworth Inc. Belmont, CA.

Salisbury, F. B., and Ross, C. W. (1985). 'Plant Physiology'. 3rd edition. Wadsworth Inc. Belmont CA.

Sangster, A. G. (1970a). Ann. Bot. **34**, 245-257.

Sangster, A. G. (1970b). Ann. Bot. **34**, 557-570.

Sangster, A. G., and Wynn Parry, D. (1976). Ann. Bot. **40**, 361-371.

Sangster, A. G., Hodson, M. J., and Wynn Parry, D. (1983). New Phytol. **93**, 105-122.

Santos, R. B., and Sutton, B. G. (1982). Aust. J. Agric. Res. **33**, 1037-1048.

Sastry, M. C. S., and Murray, D. R. (1987). J. Sci. Food Agric. **40**, 253-261.

Savage, A. D., King, J., and Gamborg, O. L. (1979). Plant Sci. Lett. **16**, 367-376.

Schmalstig, J. G., and Geiger, D. R. (1985). Plant Physiol. **79**, 237-241.

Schmalstig, J. G., and Geiger, D. R. (1987). Plant Physiol. **83**, 49-52.

Schroeder, H. E. (1984). J. Exp. Bot. **35**, 813-821.

Schubert, K. R. (1986). Annu. Rev. Plant Physiol. **37**, 539-574.

Schwartz, K. (1973). Proc. Nat. Acad. Sci. USA **70**, 1608-1612.

Scowcroft, W. R., Larkin, P. J., and Brettell, R. I. S. (1983). In 'The Use of Tissue Culture and Protoplasts in Plant Pathology' (B. J. Deverell and J. P. Helgeson, Eds), pp. 139-162. Academic Press, Sydney.

Secor, J. (1987). Plant Physiol. **83**, 143-148.

Servaites, J. C., Schrader, L. E., and Jung, D. M. (1979). Plant Physiol. **64**, 546-550.

Servaites, J. C., Geiger, D., Tucci, M. A., and Fondy, B. R. (1988). Plant Physiol. (in press).

Sesták, Z. 'Photosynthesis During Leaf Development'. Dr W. Junk Publishers, Dordrecht.

Sevilla, F., Lopez-Gorge, J., and del Río, L. A. (1982). Plant Physiol. **70**, 1321-1326.

Sevilla, F., del Río, L. A., and Hellin, E. (1984). J. Plant Physiol. **116**, 381-387.

Shanmugasundaram, S., Janardhanan, K., and Lakshmanan, K. K. (1983). Z. Pflanzenphysiol. **111**, 469-473.

Shapiro, B. M., and Stadtman, E. R. (1970). Methods in Enzymology **17**, 910-922.

Sharkey, P. J., and Pate, J. S. (1975). Planta **127**, 251-262.

Sharkey, P. J., and Pate, J. S. (1976). Planta **128**, 63-72.

Shear, C. B., and Faust, M. (1970). Plant Physiol. **45**, 670-674.

Sheldrake, A. R., and Saxena, N. P. (1979). Ann. Bot. **43**,

467-473

Shepard, J. F. (1980). In 'Genetic Improvement of Crops. Emergent Techniques' (I. Rubinstein, B. Gengenbach, R. L. Phillips and C. E. Green, Eds), pp. 185-219. University of Minnesota Press, Minneapolis U.S.A.

Shepard, J. F. (1981). Annu. Rev. Phytopathol. **19**, 145-166.

Shepard, J. F., Bidney, D., and Shahin, E. (1980). Science **208**, 17-24.

Sherrard, J. H., Kennedy, J. A., and Dalling, M. J. (1979) Plant Physiol. **64**, 439-444.

Shimamoto, K., and Nelson, O. E. (1981). Plant Physiol. **67**, 429-432.

Shrift, A. (1969). Annu. Rev. Plant Physiol. **20**, 475-494.

Simon, E. W. (1984). In 'Seed Physiology Vol. 2. Germination and Reserve Mobilization' (D. R. Murray. Ed.), pp. 77-115. Academic Press, Sydney.

Simpson, R. J., Lambers, H., and Dalling, M. J. (1983). Plant Physiol. **71**, 7-14.

Singh, B. K., and Pandey, R. K. (1980). Aust. J. Plant Physiol. **7**, 727-735.

Singh, M. B., Bhalla, P. L., and Malik, C. P. (1979). Phytomorphology **29**, 306-309.

Singh, M. B., Bhalla, P. L., and Malik, C. P. (1980a). Ann. Bot. **45**, 523-527.

Singh, M. B., Bhalla, P. L., and Malik, C. P. (1980b). Biochem. Physiol. Pflanzen **175**, 389-395.

Skoog, F., and Miller, C. O. (1957). Symp. Soc. Exp. Biol. **11**, 118-131.

Smith, F. W. (1973). 'Foliar Symptoms of Nutrient Disorders in *Chloris gayana*'. Division of Tropical Pastures Technical Paper No. **13**. CSIRO Australia.

Smith, F. W. (1978). In 'Mineral Nutrition of Legumes in Tropical and Subtropical Soils' (C. S. Andrew and E. J. Kamprath, Eds), pp. 329-346.

Smith, H., and Attridge, T. H. (1970). Phytochem. **9**, 487-495.

Smith, J. A. C., and Milburn, J. A. (1980a). Planta **148**, 28-34.

Smith, J. A. C., and Milburn, J. A. (1980b). Planta **148**, 35-41.

Smith, J. A. C., and Milburn, J. A. (1980c). Planta **148**, 42-48.

Smith, J. G. (1973). Plant Physiol. **51**, 454-458.

Snapp, S. S., and Vance, C. P. (1986). Plant Physiol. **82**, 390-395.

Snowball, K., and Robson, A. D. (1986). 'Symptoms of Nutrient Deficiencies: Lupins'. University of Western Australia Press, Perth, W.A.

Sodek, L., Lea, P. J., and Miflin, B. J. (1980). Plant Physiol. **65**, 22-26.

Solórzano, L. (1969). Limnol. Oceanog. **14**, 799-801.

Somerville, C. R. (1984). Oxford Surveys Plant Mol. Cell Biol. **1**, 103-131.

Somerville, C. R. (1986). Annu. Rev. Plant Physiol. **37**, 467-507.

Somerville, C. R., and Ogren, W. L. (1983). Proc. Nat. Acad. Sci. U.S.A. **80**, 1290-1294.

Spencer, D., and Wood, J. G. (1954). Aust. J. Biol. Sci. **7**, 425-434.

Sporne, K. R. (1954). Phytomorphology **4**, 275-278.

Spragg, S. P., and Yemm, E. W. (1954). J. Exp. Bot. **10**, 409-425.

Sprent, J. I. (1980). Plant Cell Environ. **3**, 35-43.

Stafford, A. (1978). 'Aspects of Development of the Cotyledon and Testa of *Pisum sativum*'. Ph.D. Thesis, The University of East Anglia.

Stafford, A., and Davies, D. R. (1979). Ann. Bot. **44**, 315-321.

Stewart, J. McD., and Hsu, C. L. (1977). Planta **137**, 113-117.

Stitt, M. (1987). Plant Physiol. **84**, 201-204.

Stitt, M., Mieskes, G., Soling, H.-D., and Heldt, H. W. (1982). FEBS Lett. **145**, 217-222.

Stitt, M. and Heldt, H. W. (1985). Plant Physiol. **79**, 599-608.

Storey, R., and Beevers, L. (1977). Planta **124**, 77-87.

Stout, P. R., Johnson, C. M., Carlton, A. B., and Broyer, T. C. (1954). Plant Physiol. **29**, 526-530.

Straus, J., and La Rue, C. D. (1954). Am. J. Bot. **41**, 687-694.

Street, H. E. (1954). Discovery **15**, 286-292.

Street, H. E., and Lowe, J. S. (1950). Ann. Bot. **16**, 185-205.

Streeter, J. G. (1973). Arch. Biochem. Biophys. **157**, 613-

624.

Sussex, I. M., and Dale, R. M. K. (1979). In 'The Plant Seed: Development, Preservation and Germination' (I. Rubenstein, R. L. Phillips, C. E. Green and B. G. Gengenbach, Eds), pp. 129-141. Academic Press, N.Y.

Swamy, B. G. L., and Ganapathy, P. M. (1959). Bot. Gaz. **119**, 47-50.

Swanson, C. A. (1959). In 'Plant Physiology Vol. II. Plants in Relation to Water and Solutes' (F. C. Steward, Ed.), pp. 481-551. Academic Press, N.Y.

Ta, T. C., and Joy, K. W. (1986). Planta **169**, 117-122.

Ta, T. C., Joy, K. W., and Ireland, R. J. (1984a). Plant Physiol. **74**, 822-826.

Ta, T. C., Joy, K. W., and Ireland, R. J. (1984b). Plant Physiol. **75**, 527-530.

Tan, S. C. (1980). Aust. J. Plant Physiol. **7**, 159-167.

Thimann, K. V. (1987). In 'Plant Senescence: Its Biochemistry and Physiology' (W. W. Thomson, E. A. Nothnegel and R. C. Huffaker, Eds). University of California.

Thomas, S., and Corden, M. (1970). 'Tables of Composition of Australian Foods'. Aust. Government Pub. Service, Canberra, A.C.T.

Thompson, J. F., Madison, J. T., and Muenster, A. E. (1977). Ann. Bot. **41**, 29-39.

Thompson, J. F., Madison, J. T., Waterman, M. A., and Muenster, A. E. (1981). Phytochem. **20**, 941-945.

Thorne, J. H. (1979). Agron. J. **71**, 812-816.

Thorne, J. H. (1980). Plant Physiol. **65**, 975-979.

Thorne, J. H. (1981). Plant Physiol. **67**, 1016-1025.

Thorne, J. H. (1982a). Plant Physiol. **69**, 48-53.

Thorne, J. H. (1982b). Plant Physiol. **70**, 953-958.

Thorne, J. H. (1985). Annu. Rev. Plant Physiol. **36**, 317-343.

Thorne, J. H., and Giaquinta, R. T. (1984). In 'Storage Carbohydrates in Vascular Plants' (D. H. Lewis, Ed.), pp. 75-96.

Thorne, J. H., and Rainbird, R. M. (1983). Plant Physiol. **72**, 268-271.

Ting, I. P. (1985). Annu. Rev. Plant Physiol. **36**, 595-622.

Ting, I. P., and Osmond, C. B. (1973a). Plant Physiol. **51**, 439-447.

Ting, I. P., and Osmond, C. B. (1973b). Plant Physiol.

51, 448-453.

Todd, G. W., Bean, R. C., and Propst, B. (1961). Plant Physiol. **36**, 69-73.

Tolbert, N. E. (1973). In 'Current Topics in Cellular Regulation' (B. L. Horecker and E. R. Stadtman, Eds), pp. 21-50. Academic Press, London, N.Y.

Tolbert, N. E., Oeser, A., Kisaki, T., Hageman, R. H., and Yamazaki, R. K. (1968). J. Biol. Chem. **243**, 5179-5184.

Tolbert, N. E., Yamazaki, R. K., and Oeser, A. (1970). J. Biol. Chem. **245**, 5129-5136.

Tran, V. N., and Cavanagh, A. K. (1984). In 'Seed Physiology Vol. 2. Germination and Reserve Mobilization' (D. R. Murray, Ed.), pp. 1-44. Academic Press, Sydney.

Tromp, J., and Ovaa, J. C. (1971). Physiol. Plant. **25**, 16-22.

Tully, R. E., and Hanson, A. D. (1979). Plant Physiol. **64**, 460-466.

Tunstall, B. R., and Connor, D. J. (1975). Aust. J. Plant Physiol. **2**, 489-499.

Turner, D. H., and Turner, J. F. (1957). Aust. J. Biol. Sci. **10**, 302-309.

Turner, D. H., and Turner, J. F. (1960). Biochem. J. **74**, 486-491.

Turner, J. F., Turner, D. H., and Lee, J. B. (1957). Aust. J. Biol. Sci. **10**, 407-413.

Umbeck, P. F., and Gengenbach, B. G. (1983). Crop Sci. **23**, 584-588.

Urquhart, A. A., and Joy, K. W. (1981). Plant Physiol. **68**, 750-754.

Usuda, H., and Edwards, G. E. (1980). Plant Physiol. **65**, 1017-1022.

Vairinhos, F., and Murray, D. R. (1982). Z. Pflanzenphysiol. **106**, 447-452.

Van Bel, A. J. E. (1984). Plant Sci. Lett. **35**, 81-85.

Van Bel, A. J. E., and Hermans, H. P. (1977). Z. Pflanzenphysiol. **84**, 413-418.

Van Bel, A. J. E., and Patrick, J. W. (1984). Plant Growth Reg. **2**, 319-326.

Van Bel, A. J. E., and Patrick, J. W. (1985). Plant Cell Environ. **8**, 1-8.

Van Beusichem, M. L. (1983). Z. Pflanzenphysiol. **109**, 449-458.

Vandenbelt, J. M. (1945). Nature **156**, 175.

Van Die, J. (1974). Acta Bot. Neerl. **23**, 521-540.

Van Die, J., and Willemse, N. C. M. (1975). **24**, 237-239.

Van Die, J., and Willemse, N. C. M. (1980). Ber. Deutsch. Bot. Ges. **93**, 327-337.

Van Goor, B. J., and Wiersma, D. (1976). Physiol. Plant. **36**, 213-216.

Van Overbeek, J., Conklin, M. E., and Blakeslee, J. (1942) Am. J. Bot. **29**, 472-477.

Verin-Vergeau, C., Baldy, P., and Cavalié, G. (1979). Phytochem. **18**, 1279-1282.

Verma, D. C., and Dougall, D. K. (1977). Plant Physiol. **59**, 81-85.

Ver Nooy, C. D., Thorne, J. H., Lin, W., and Rainbird, R. M. Plant Physiol. **82**, 222-225.

Vose, P. B. (1963). J. Exp. Bot. **14**, 448-457.

Vose, P. B. (1980). 'Introduction to Nuclear Techniques in Agronomy and Plant Biology'. Pergamon Press, Oxford.

Wagner, G., and Klein, K. (1978). Z. Pflanzenphysiol. **79**, 283-291.

Walker, C. D., and Loneragan, J. F. (1981). Ann. Bot. **47**, 65-73.

Walker, D. A., and Sivak, M. N. (1985). Physiol. Vég. **23**, 829-841.

Walker, D. A., and Sivak, M. N. (1986). TIBS **11**, 176-179.

Wallace, W. (1973). Plant Physiol. **52**, 197-201.

Wallace, W. (1975). Plant Physiol. **55**, 774-777.

Wallace, W. (1986). Physiol. Plant. **66**, 630-636.

Wallace, W., and Pate, J. S. (1965). Ann. Bot. **29**, 655-671.

Wallace, W., and Pate, J. S. (1967). Ann. Bot. **31**, 213-228.

Wallsgrove, R. M., Turner, J. C., Hall, N. P., Kendall, A. C., and Bright, S. W. J. (1987). Plant Physiol. **83**, 155-158.

Walton, N. J., and Woolhouse, H. W. (1986). Planta **167**, 119-128.

Wara-Aswapati, O., and Bradbeer, J. W. (1974). Plant Physiol. **53**, 691-693.

Wardlaw, C. W. (1952). 'Phylogeny and Morphogenesis'. MacMillan and Co. Ltd., London.

Wardlaw, I. F. (1974). Annu. Rev. Plant Physiol. **25**, 515-

539.

Wardlaw, I. F. (1982). Aust. J. Plant Physiol. **9**, 705-713.

Warembourg, F. R., and Montagne, D. (1982). Physiol. Plant. **56**, 46-55.

Waters, S. P., Peoples, M. B., Simpson, R. J., and Dalling M. J. (1980). Planta **148**, 422-428.

Waters, S. P., Noble, E. R., and Dalling, M. J. (1982). Plant Physiol. **69**, 575-579.

White, P. R. (1943). Growth **7**, 53-65.

Williams, M., and Randall, D. D. (1979). Plant Physiol. **64**, 1099-1103.

Willmer, C. M. (1983). 'Stomata'. Longman, London and N.Y.

Wills, R. B. H., Scriven, F. M., and Greenfield, H. (1983). J. Sci. Food Agric. **34**, 1383-1389.

Wills, R. B. H., Evans, T. J., Lim, J. S. K., Scriven, F. M., and Greenfield, H. (1984). Food Technol. Aust. **36**, 512-514.

Winkler, R. G., Polacco, J. C., Eskew, R. M., and Welch, R. M. (1983). Plant Physiol. **72**, 262-263.

Wolswinkel, P. (1985a). J. Plant Physiol. **120**, 419-429.

Wolswinkel, P. (1985b). Physiol. Plant. **65**, 331-339.

Wolswinkel, P. (1987). Plant Physiol. Biochem. **25**, 557-566.

Wolswinkel, P., and Ammerlaan, A. (1983). Planta **158**, 205-215.

Wolswinkel, P., and Ammerlaan, A. (1984). Physiol. Plant. **61**, 172-182.

Wolswinkel, P., and Ammerlaan, A. (1985a). J. Exp. Bot. **36**, 359-368.

Wolswinkel, P., and Ammerlaan, A. (1985b). Ann. Bot. **56**, 35-43.

Wolswinkel, P., and Ammerlaan, A. (1986). Plant Cell Environ. **9**, 133-140.

Wolswinkel, P., and de Ruiter, H. (1985). Ann. Bot. **55**, 283-287.

Wolswinkel, P., Ammerlaan, A., and Kuyven Hoven, H. (1983) Physiol. Plant. **59**, 375-386.

Wolswinkel, P., Kraus, E., and Ammerlaan, A. (1986). J. Exp. Bot. **37**, 1462-1471.

Wolterbeek, H. Th., and de Bruin, M. (1986a). J. Exp. Bot. **37**, 331-340.

Wolterbeek, H. Th., and de Bruin, M. (1986b). J. Exp. Bot. **37**, 928-939.

Wolterbeek, H. Th., van Luipen, J., and de Bruin, M. (1984). Physiol. Plant. **61**, 599-606.

Woodrow, I. E., and Rowan, K. S. (1979). Aust. J. Plant Physiol. **6**, 39-46.

Wright, J. P., and Fisher, D. B. (1983). Plant Physiol. **73**, 1042-1047.

Yeung, E. C. (1983). Protoplasma **118**, 225-229.

Yeung, E. C., and Brown, D. C. W. (1982). Z. Pflanzenphysiol. **106**, 149-156.

Yeung, E. C., and Sussex, I. M. (1979). Z. Pflanzenphysiol. **91**, 423-433.

Young, C. T., and Mason, M. E. (1972). J. Food Sci. **37**, 722-725.

Yu, C., Liang, Z., and Huang, A. H. C. (1984). Plant Physiol. **75**. 7-12.

Yu, Y. L., Tolbert, N. E., and Orth, G. M. (1964). Plant Physiol. **39**, 643-647.

Zeng, J. Z. (1983). In 'Plant Cell Culture in Crop Improvement' S. K. Sen and K. L. Giles, Eds), pp. 351-363. Plenum Press, N.Y.

Zimmermann, M. H. (1960). Annu. Rev. Plant Physiol. **11**, 167-190.

Zimmermann, M. H. (1969). In 'The Physiology of Plant Growth and Development' (M. B. Wilkins, Ed.), pp. 383-417. McGraw-Hill, Maidenhead, U.K.

Zimmermann, M. H., and Milburn, J. A. (1975). Encyclopaedia of Plant Physiology Vol. 1. Springer-Verlag, Berlin.

Addenda

Bhardwaj, S. N., and Raju, T. V. K. (1972). Ind. J. Plant Physiol. **15**, 38-55.

Chen, J., and Kennedy, I. R. (1985). Phytochem. **24**, 2167-2172.

Eeuwens, C. J., and Schwabe, W. W. (1975). J. Exp. Bot. **26**, 1-14.

Hartman, C. L., McCoy, T. J., and Knous, T. R. (1984a). Plant Sci. Lett. **34**, 183-194.

Hartman, C. L., Knous, T. R., and McCoy, T. J. (1984b). Phytopathology **74**, 818.

Hawker, J. S., and Buttrose, M. S. (1980). Ann. Bot. **46**, 313-321.

Maranville, J. W. (1970). Plant Physiol. **45**, 591-593.

Marinos, N. G. (1970). Protoplasma **70**, 261-279.

Millerd, A., Spencer, D., Dudman, W. F., and Stiller, M. (1975). Aust. J. Plant Physiol. **2**, 51-59.

Murray, D. R., and Bradbeer, J. W. (1971). Phytochem. **10**, 1999-2003.

Murray, D. R., and Knox, R. B. (1977). J. Cell Sci. **26**, 9-18.

Norstog, K., and Smith, J. E. (1963). Science **142**, 1655-1656.

Tachini, P., and Walbot, V. (1987). Nestlé Research News 1986/87, 18-29.

Timothy, D. H., Levings, C. S. III, Pring, D. R., Conde, M. F., and Kermicle, J. L. (1979). Proc. Nat. Acad. Sci. U.S.A. **76**, 4220-4224.

Tsai, H. S., and Tseng, M. T. (1979). Bot. Bull. Academica Sinica **20**, 117-122.

Wallace, W., Secor, J., and Schrader, L. E. (1984). Plant Physiol. **75**, 170-175.

Wetherell, D. F., and Dougall, D. K. (1976). Physiol. Plant. **37**, 97-103.

Index

B

C

O

P

Q

R

S

T

U

V